79.00

THE ASIC HANDBOOK

Nigel Horspool
Peter Gorman

PRENTICE HALL PTR
UPPER SADDLE RIVER, NJ 07458
WWW.PHPTR.COM

ISBN 0-13-091558-0

9 780130 915580

90000

Library of Congress Cataloging-in-Publication Data

Horspool, Nigel

   The ASIC Handbook / Nigel Horspool and Peter Gorman

      p.   cm,

   ISBN 0-13-091558-0

   1. Application specific integrated circuits.   I. Gorman, Peter.   II. Title.

   TK7874.6.H67 2001

   621.39'5--dc21

                                                                20-01021426

Editorial/production supervision: *Nicholas Radhuber*
Cover design: *Anthony Gemmellaro*
Cover design director: *Jerry Votta*
Manufacturing manager: *Alexis R. Heydt*
Marketing manager: *Dan DePasquale*
Acquisitions editor: *Bernard Goodwin*
Editorial assistant: *Michelle Vincenti*
Composition: *Wil Mara*

© 2001 Prentice Hall PTR
Prentice-Hall, Inc.
Upper Saddle River, NJ 07458

The publisher offers discounts on this book when ordered in bulk quantities. For more information contact: Corporate Sales Department, Prentice Hall PTR, One Lake Street, Upper Saddle River, NJ 07458. Phone: 800-382-3419; Fax: 201-236-7141; E-mail: corpsales@prenhall.com.

Printed in the United States of America

10 9 8 7 6 5 4 3 2 1

ISBN 0-13-091558-0

Prentice-Hall International (UK) Limited, *London*
Prentice-Hall of Australia Pty. Limited, *Sydney*
Prentice-Hall Canada Inc., *Toronto*
Prentice-Hall Hispanoamericana, S.A., *Mexico*
Prentice-Hall of India Private Limited, *New Delhi*
Prentice-Hall of Japan, Inc., *Tokyo*
Pearson Education Asia Pte. Ltd.
Editora Prentice-Hall do Brasil, Ltda., *Rio de Janeiro*

To our families:
Jane, Catherine, and Rachel Horspool,
and Marie Clare Gorman
for their patience and support

CONTENTS

Chapter 3 • A Quality Design Approach   53

Chapter 4 • Tips and Guidelines   69

PREFACE

This book is a practical, step-by-step guide to the process of designing digital Application-Specific Integrated Circuits, or ASICs, as they are universally referred to in the industry. These components lie at the heart of nearly all successful electronic products. In the early 1990s, only a relatively small number of companies had in-house ASIC design teams. Outside of these, third-party ASIC design companies serviced the rest of what was still a relatively small market. ASIC know-how was considered an esoteric subject. By the late 1990s, less than 10 years later, this situation had transformed far beyond what anyone could have projected. Access to ASIC expertise had become and remains a survival requirement for all the major companies in the electronics industry and for many small and medium-sized enterprises, too. Such has been the explosive growth in demand for experienced ASIC teams that there is now a significant shortfall in supply. Those companies that do succeed in attracting ASIC expertise and developing it to its maximum potential hold the key to making market-winning products that can yield enormous returns on investment. Herein lies the value of this publication.

The book's aim is to highlight all the complex issues, tasks and skills that must be mastered by an ASIC design team in order to achieve successful project results. It targets ASIC and non-ASIC readers in its scope. The techniques and methodologies prescribed in the book, if properly employed, can significantly reduce the time it takes to convert initial ideas and concepts into right-first-time silicon. Reducing this ever-critical time to market does not simply save on development costs. For new products or new market segments, it provides the opportunity for getting the product there ahead of the competition and, thus, creates the potential for significantly increased market share.

The book covers all aspects of ASIC-based development projects. It includes a detailed overview of the main phases of an ASIC project. Dedicated chapters provide comprehensive coverage of the key technical issues, and a further section of the book deals with relevant management techniques. The technical methods include design for reuse, high-quality design approaches, VHDL/Verilog coding tips and synthesis guidelines. Management skills such as team building are presented, as are ASIC leader tasks such as planning, risk reduction and managing relationships with ASIC vendors.

The book has been written by two ASIC consultants who have worked on many successful ASIC projects in a variety of companies. They are interested in both the technical and management aspects of ASIC design. They are motivated by a desire to find and formulate continuous improvements in approaches to design and development processes. The book was written partly for their own benefit, to capture their own experiences with a view to helping them reproduce successful techniques and methodologies on future projects. Their hope now is that others can also benefit from their work. The book is intended to act as a companion guide to an ASIC team. It can be read in its entirety or subject by subject, as the need arises. It should be reread at the outset of each project and referred to frequently as the project progresses.

Who Should Read This Book

The book is aimed at anyone who wants to understand the elements of an ASIC project. It is also aimed at anyone interested in improving quality, reducing risks and improving time to market. Although some prior knowledge is an advantage in reading some of the more technical chapters, many sections of the book can be read and understood by beginners; therefore, the book is a good starting point for people beginning ASIC careers or contemplating this as an option. Broadly speaking, three groups of people will be interested in the book: non-ASIC engineers, ASIC project managers and ASIC design engineers.

Non-ASIC People

There are many people who have a vested interest in ASIC projects achieving their goals. The ASIC often forms part of a larger overall project that combines software, printed circuit board, mechanical and system design disciplines. In a multi-disciplined team, it is essential that at least a number of key people have a degree of cross-discipline knowledge. Such knowledge enables them to understand the opportunites and limitations that exist when the various disciplines come together, and it allows them to make educated decisions and trade-offs. Additionally, most progressive engineers have a natural desire to learn about other disciplines on their own initiative, whether it is an absolute requirement or not. This book provides the non-ASIC engineer with the opportunity to understand the world of ASICs.

Senior managers, many of whom predate the ASIC revolution or come from completely different backgrounds, such as marketing or finance, also stand to benefit from this book. In addition to the management essentials that are covered, sections from several of the technical chapters will be within their grasp and will add to their knowledge of what they are dealing with.

ASIC Project Managers

ASIC projects are very complex and require project managers with a wide range of knowledge and skills. Project managers are frequently promoted from design engineering positions, where they may have been focused on relatively narrow or specialist areas. They are, therefore, often thrown in at the deep end when it comes to knowledge of the wider design process and knowledge of management techniques. This book is aimed at those project managers who want to understand and improve the entire ASIC design process. The book clearly explains the project flow and quality approaches to design. It is also useful for project managers who want to improve their project management skills covering issues such as team motivation, managing third-party ASIC vendors and monitoring and managing risks through all the phases of a project.

Design Engineers

Good design engineers are always keen to improve their knowledge and the quality of their work. They may initially participate in only a limited part of the ASIC design process, and it may take several years before they get hands-on exposure to the full spectrum of activities involved. This book defines the full ASIC process, describing good design practices, guidelines for reuse, top-down methodologies and coding and synthesis approaches. The design techniques described will enable engineers to design to a higher quality in shorter time scales.

The Structure of the Book

The book is structured into a number of parts that are formed from one or more chapters. The first chapter, "Phases of an ASIC Project," in section 1, "Project Overview," gives a detailed description of each of the phases of an ASIC project. It provides an overview of technical issues and planning tasks that are required at each stage of the design.

The second section, "Design Techniques," deals with technical issues, including design reuse, quality design approaches, simulation techniques, VHDL/Verilog coding tips, and synthesis guidelines. These chapters provide practical advice on these topics. They address higher-level problems, such as design approach and quality test environments, rather than presenting an academic course on low-level device physics. They are useful for both seasoned and less experienced design engineers.

The next section, "Project Management," deals with project management aspects such as planning, risk reduction and dealing with ASIC vendors. The techniques described here can be reapplied to successive projects and refined to match best the characteristics of each new project that is undertaken. Although there are numerous books available on project management, they tend to be quite broad in their scope. The project management section in this book treats the subject in an ASIC-specific context.

The fourth section, "People and Team Management," aims to provide the project manager with an introduction to the subject of people and team management. Again, this is framed in an ASIC delevopment context. Some basic knowledge of motivation techniques, communication issues and team management theory will help projects to run more smoothly. A well-structured and highly motivated team will bring a project to a successful conclusion sooner and produce quality results that can be reused in future designs.

The book closes with "EDA Tools." There is an abundance of EDA tools on the market. This chapter picks out a few of the more commonly used types and explains their purpose.

ACKNOWLEDGMENTS

We would like to thank the following people, who have contributed in a variety of ways toward making this book possible—David Law, Killian O'Sullivan, James Wright, Frank Werner, David Wilton, Charles Dancak, Jim Hendron and Teresa Lennon.

## CHAPTER 1

Phases of an ASIC Project

1.1 Introduction

This chapter provides an overview of the entire Application-Specific Integrated Ciruit (ASIC) development process. Starting from feasibility, it breaks the process down into a set of project phases that must be carried out to obtain fully validated silicon that can be used by production. The main activities in each phase are described, as are the roles and responsibilities of the different members of the project team during these phases. Because this chapter is intended to serve as an overview, the reader is referred to subsequent chapters in the book for detailed treatment of many of the issues covered here.

1.2 List of Phases

Different companies and design teams may use slightly different names to describe what is essentially the same set of steps in the ASIC design process. The phases listed below, or slight variations on them, are followed by most ASIC design teams today. The names or titles used for each phase are intended to capture the main activities during the phase. However, within each phase, there are always miscellaneous tasks that do not fit neatly under the title associated with the main activity. These miscellaneous tasks are, of course, still described because they are all necessary activities that take place during the phase in question.

The main phases of an ASIC project are:

- Top-Level Design
- Module Specification
- Module Implementation
- Subsystem Simulation
- System Simulation and Synthesis
- Layout and Backend
- Preparation for Testing of the Silicon
- ASIC Sign-Off
- Testing of the Silicon

The phases are shown diagrammatically in Figure 1-1.

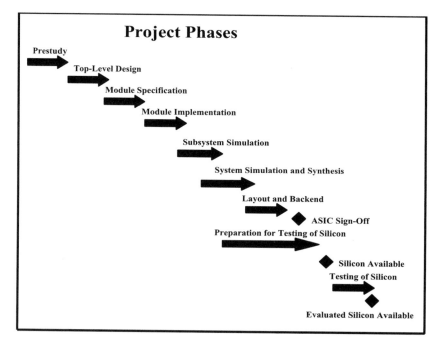

Figure 1-1   List of Project Phases

The following subsections describe the phases of the project in greater detail. The phases are generally sequential in time, although there can be some element of overlap. The initial prestudy, top-level design and specification stages have a crucial impact on the chances of success of the entire project. It is, therefore, important to allow adequate time and appropriate resources for the initial phases and to avoid the tendency to rush too quickly into the implementation phase.

1.3 Prestudy Phase

Outputs from This Phase:

- Estimate of project timescales and resource requirements
- Estimate of silicon area
- Estimate of product cost
- Initial architecture design
- Analysis of risks
- Identification of project design goals, deliverables and milestones
- Initial decisions on design route and tools

General Tasks:

- Initial architecture design
- Initial planning and estimation of resource requirements
- Risk and cost analysis

The prestudy phase is the initial stage of the project where development and marketing work closely together. The prestudy work must identify a business opportunity and some initial product architectures to satisfy the opportunity. The opportunity may arise as a result of a new market area or a new design to reduce costs or add new system features. Often, ASIC projects integrate a number of functions into one chip, reducing cost and power and typically increasing performance and providing additional features. This is a key characteristic of the emerging SoC (system-on-a-chip) market, where entire systems are integrated onto one chip, typically by reusing a combination of in-house and bought-in third-party IP (intellectual property).

The deliverable from the prestudy phase will be a business case that projects the financial returns, development timescales and risks. The development part of the business case should provide estimates for product costs, timescales and resources needed. It should also provide a detailed assessment of the risks.

If the purpose of the ASIC is to replace a currently successful product, cost and/or feature enhancement is often the driving requirement. When the purpose is to address a new market area or to replace a struggling product, timescales are often the highest priority. The project leader should establish the driving factors before starting the prestudy phase, because these will affect the architecture options.

The project leader will usually require resources during the prestudy phase. These should include experienced engineers who are good at top-level design. It is important to have access to both hardware and software resources. This is a good time to build the core of a strong team. The team that works on the prestudy phase should ideally be involved throughout the full duration of the project and especially during the top-level and design phases.

During this prestudy phase, the ASIC team must identify the major building blocks of the ASIC and obtain some initial ASIC cost quotations. This will involve talking to ASIC vendors to get a current range of prices. It is useful to consider a number of product solutions and architectures that differ in terms of cost, development timescales, resource requirements, etc. A set of options can then be presented to the management team with a recommended option identified. The team should justify why a particular option is recommended.

For SoC designs or very large devices, any external third-party IP block requirements must be identified and initial cost quotations obtained. If any such IP blocks are not already predeveloped by the IP provider, development timescales and delivery dates must be determined. The team should be wary of selecting IP that has not been proven on silicon because it presents an obvious additional risk to the project.

Large, deep submicron designs may require the use and purchase of new EDA (electronic design automation) tools. Any such requirements should be identified and discussed at this stage with the ASIC vendor and the IP provider because the costs may be significant and should, therefore, be incorporated in the business case.

As part of the process of identifying product solutions, the team should weigh the merits of an ASIC-based solution against the possibility of using third-party silicon. The ASIC may be cheaper or may provide important features or a performance advantage. It must provide some marketable benefit. Otherwise, it is pointless to embark on an ASIC development.

There are no major risks during the prestudy phase. As always, however, there are two conflicting pressures; the first is to generate an accurate business plan defining a winning product, and the second is the pressure of timescales. The prestudy phase can take a reasonable length of time. In this phase, product innovation is very important. Defining the right product and the right development approach makes a winning product. If the prestudy phase is rushed, the chances of success are reduced.

The project leader can reduce the risk of lengthy prestudy timescales by requesting appropriate resource in a timely manner and having all relevant information available when the management team assesses the business case.

Good product or development ideas can be generated when a number of engineers brainstorm ideas together. Several brainstorming sessions should be held. These should include both hardware and software representatives, and participation should not be limited exclusively to those working full time on the prestudy. It can prove beneficial to draft additional contributors because of their known expertise in the area or because of their known capacity for innovation and creative thinking. Competing products should be compared with the product ideas.

1.4 Top-Level Design Phase

Outputs from This Phase:

- Functional requirements specification reviewed and agreed on.
- Top-level architecture document reviewed and agreed on.
- Initial plan and resource requirements reviewed and agreed on.

General Tasks:

- Write functional requirements specification.
- Brainstorm a number of architectural options.
- Analyze architecture options—consider technical feasibility, resource requirements, development timescales, etc.
- Generate top-level architecture document.
- Identify critical modules—start them early, if necessary.
- Identify possible third-party IP block requirements.
- Select team members.
- Identify new processes and tools.
- Agree design route/design flow.
- Brainstorm risks.
- Estimate silicon area, pin-out, cost, power, etc.

Project Manager-Specific Tasks:

- Generate project plan.
- Obtain resources (project team, equipment, tools).
- Organize training courses.

This is a creative phase, during which the top-level product architecture is defined.

Many of the classic engineering trade-offs are made in this phase. Factors such as cost of the product, cost of the design, time to market, resource requirements and risk are compared with each other as part of the process of developing the top-level design. Innovation at this time can have dramatic effects on the success of the product. The innovation can take the form of product ideas, top-level architecture ideas and design process ideas. The resource requirements during this stage will again be a small number of skilled engineers who are talented architects and system designers.

The team should make trade-offs between functions performed by standard off-the-shelf chips, existing ASICs, software and the new ASIC. Chapter 8, "Planning and Tracking ASIC Projects," outlines some ideas for reducing timescales through careful architectural design.

One of the deliverables at the end of this stage is a top-level architecture document, which clearly defines the partition between board, software and ASIC functions. Often, the ASIC represents such a significant part of the design that the top-level ASIC functions are also defined in the architecture specification. It is important to review the top-level architecture specification with selected experts within the company, including representatives from other project teams.

At this stage, an initial ASIC plan should be developed. This plan and the resource requirements should be reviewed and agreed with senior management. At the same time, the design route document should be written. This defines the tools, techniques and methodologies that will be used at each stage. It will aid the planning process because it forces the team to think about required tools and equipment early on. It may be necessary to start booking these at this stage. From the point of view of the design team, the design route document provides a good overview of the design flow and explains the reasons for some of the project rules. The design route document is described in Chapter 3, "A Quality Design Approach." Any new tool identified in the design route document should be tested before the tool is needed, and the plan should include time for testing and contingency options if any of the tools are not sufficiently reliable or simply do not work.

Toward the end of this stage, the team members are selected. The project manager should identify any training requirements and ensure that courses are booked. It is helpful at this stage to assign an engineer familiar with the ASIC architecture to prepare a presentation explaining the architecture. This can then be used to get new team members up to speed as they join the project.

During this phase of the project, many factors will be unknown. The project leader must make educated guesses about resources, timescales, costs, risks and what the competitors are likely to do. Modules, which are critical from the point of view of timescale or risk, should be identified and, if necessary, work should be started on these early. One method of reducing risk and improving timescales is to consider design reuse (see Chapter 2, "Design Reuse and System-on-a-Chip Designs," for details).

For SoC designs or very large designs, IP reuse is a common design approach. It may be necessary to source IP blocks from third-party vendors. Such large designs will typically require significant effort during the top-level design phase because the size of the chip often implies a high level of complexity. One or more team members should be assigned the task of identifying IP blocks from other companies, analyzing their specifications and assessing the impact on the chip architecture. This task should actually begin in the prestudy phase, but the level of detailed analysis is greater during this phase. The project manager should begin contacting IP providers to start negotiating price, deliverables and support. Some of the issues associated with the use of external third-party IP are considered in Chapter 2, "Design Reuse and System-on-a-Chip Designs."

Equally important, existing IP blocks from within the company should be analyzed. The advantages of using in-house IP are lower cost, in-house expert support and the possibility of making minor modifications to the IP (although, wherever possible, reusable blocks should not be modified). The disadvantage associated with in-house IP blocks is that if they were not originally designed and documented with reuse in mind, they may not facilitate easy reuse. In extreme cases, reusing blocks not designed with reuse in mind can take longer than redesigning them from scratch. In less extreme cases, when the blocks are being used in a different environment or in a different way than before, they may exhibit previously undiscovered bugs. The project manager should ensure that either sufficient in-house expert support is available for the project or, alternatively, that the design has been certified as a reusable block that is compliant with the company's reuse standards.

For projects that are using deep submicron technology, the design route should be identified at this stage. Any new tools required should be sourced and tested, and any necessary training programs on the use of these tools should be organized.

During the top-level design phase, there is a risk of moving too quickly to the next stage, before the top-level design document and a top-level plan are available. The next phase of the project is when the majority of the team comes on board. Time can be wasted when many people join a project if the top-level architecture is not properly defined and documented.

Toward the end of the top-level design phase, it is useful to start thinking about which engineers will work on which blocks. This can be particularly useful for the less experienced engineers. They can be overwhelmed by the system in its entirety and may feel more comfortable focusing at an early stage on specific parts of the design.

When the resource requirements are not available within the organization, the project manager must hire new staff. This recruitment process should start early in this phase because there is usually a significant lead time in recruiting good engineers. Chapter 13, "Project Manager Skills," provides some guidelines on interviewing.

1.5 Module Specification Phase

Outputs from This Phase:

- All lower-level modules specified
- Accurate project plan available and reviewed

Tasks:

- Decompose architecture into lower-level modules.
- Document module functions and interfaces.
- Review project plan and top-level architecture document.
- Analyze risks—modify architecture/plan to reduce risks, if necessary.
- Make team aware of any standard approaches (coding style, directory structure, synthesis scripts, etc).
- Check chip design rules (die temperature, package cavity, power pin requirements, etc).
- Re-estimate silicon gate count.

Project Manager-Specific Tasks:

- Start team building activities and individual motivation techniques
- Analyze and manage risks.
- Update the plan, assigning resource to the lower-level module design tasks.
- Start to think about silicon evaluation/validation.
- Define a quality project framework that explains what information is stored where and how updates are controlled.

Risks:

- Some team members can feel isolated during the architecture design.
- The team may not understand the project goals.

During this phase, the majority of the team joins the project. It is convenient to split the task descriptions into two sections. Tasks for the bulk of the team are described first. This is followed by a summary of the other tasks that are also part of this phase.

1.5.1 Tasks for the Bulk of the Team

During this phase, the top-level architecture is decomposed and partitioned into smaller modules, and the interfaces between the modules are defined and documented. Ideally, the hierarchy of the architecture should be captured diagrammatically, using a suitable graphical tool. Some such tools can convert the captured diagrams into structural VHDL or Verilog.

It is useful to have the entire team participate during this phase. First, the top-level architecture should be presented and explained. Then the ASIC is decomposed into smaller and smaller blocks. Each block should have its interfaces defined and its function clearly stated. This process helps ensure that the entire team has a reasonable understanding of the overall design and, in particular that they have a good understanding of the specific blocks for which they have design responsibility and the blocks that interface directly to their blocks. This process encourages team spirit and motivation because the entire team is assembled and everyone gets an overview of the project. One of the deliverables at the end of this phase is to start building team spirit and motivation.

The initial architecture design meetings may take several days (depending on the complexity of the ASIC). The main objective of the initial meetings is to develop and document the lower-level architectural details. This provides a baseline from which the architecture can be refined and developed. The process normally works best if it is broken into two parts. First, there is an initial design session with everyone present. Then, after spending some time documenting and analyzing the initial work, a second meeting is held to review and further refine the architecture.

The architecture hierarchy diagram is not a static document at this stage, and care should be taken to keep it up to date. After the initial sessions, the team can split into subgroups to develop the architectures for the submodules independently. The system architect should attend as many design sessions as possible.

With the breakdown of the architecture, the timescales for the submodules can be estimated and the team members assigned to develop specific modules.

The architecture sessions usually involve defining the modules and their interconnections on a white board in a form of schematic capture process. The project leader should try to involve all the engineers in this process, rather than let it be dominated by a handful of senior team members. Junior team members may have valuable contributions to make. Even if they do not, keeping them involved improves the learning process for them. Every team member is important and should be made to feel part of the team.

1.5.2 Other Tasks

During this phase, the ASIC vendor should be selected. Chapter 10, "Dealing with the ASIC Vendor," includes advice on selecting vendors. Regular meetings should be established with the selected vendor, and the ASIC architecture and design route should be discussed. Particular attention should be paid to sign-off requirements and tools, especially if the vendor or the sign-off tools are new. This is important so that testing of the vendor tools can be added to the plan.

For SoC or large designs that use IP blocks from external companies, contracts should be signed with the IP providers during this phase. Some initial testing of these blocks should be done and analysis of simulation times, programming interfaces and test coverage used to provide input to testbench generation and simulation planning.

The project manager should be considering techniques for derisking the project and reducing timescales. These topics are discussed in Chapter 9, "Reducing Project Risks." This is the stage of the project where ideas for risk reduction and improving timescales must be generated and analyzed. Some ideas may require further investigation. These investigative tasks should be planned right away so that the ideas can be evaluated and any decisions dependent on the evaluation made as early as possible.

One of the key roles of the project manager is to build team spirit and motivation. There are many techniques that the project manager can employ, and these are discussed in Chapter 11, "Motivation and People Management." The project manager should explain the importance of the project for the business and ensure that the contribution of each engineer is valued. An important technique that helps sustain motivation among team members is the provision of relevant training programs throughout the project. One form of training that is beneficial and relevant to the project is to organize presentations by senior engineers on subjects such as design techniques, system partitioning, tools, etc.

At this stage in the project, the team members must be allocated roles and responsibility for design modules (see Chapter 12, "The Team," for a description of the different roles in the team). These roles should be clearly described to the team. It is a good idea to discuss the role allocation with some of the senior engineers. Then, following an initial assignment of roles resulting from these discussions, each individual engineer should have his or her role explained before making it public. This will strengthen the relationship between the project leader and the senior engineers, and more than likely will result in better engineer role-matching than would be the case if the project leader assigns the roles without consultation.

One of the key roles in the project is that of the testbench engineer. During this phase, the testbench top-level objectives and architecture must be defined. The testbench engineer should also attend the architecture design sessions because this will add to his or her understanding and result in a better testbench. It will also help make testbench engineers feel part of the team, because they tend to work more in isolation than those working directly on the chip design itself.

1.6 Module Design

Outputs from This Phase:

- All modules designed, coded and tested at module level (with reviews)
- Initial trial synthesis of each module
- Agreed pin-out

Tasks:

- Module design, coding, testing and synthesis
- Chip-level testbench design, coding and testing
- Generation of a more accurate silicon area estimate

Project Manager-Specific Tasks:

- Provide documentation guidelines and set expectations on standards and levels of documentation.
- Explain the reviewing process and identify what will be reviewed and when.
- Review the quality framework approach with the team.
- Run weekly project meetings, tracking and closing actions continuously.
- Agree on an approach for trial layouts with the vendor.
- Agree the test vector approach and required test coverage.
- Book resources for prototyping and testing the silicon.
- Source third-party simulation models.

Risks:

- The timescales can slip during this stage—hold early reviews, monitor plan.

- The silicon gate count may exceed maximum estimates—consider modifications to architecture.

During this phase, most of the team members are assigned to address a number of mainstream tasks, while the rest of the team undertakes the remaining miscellaneous tasks. Again, it is convenient to subdivide the phase into two activity streams. Tasks for the bulk of the team are described first, then the remaining tasks are covered.

1.6.1 Tasks for the Bulk of the Team

The architecture design will have partitioned the ASIC into a number of submodules. In some cases, the submodules will be small enough for a single engineer to design, whereas in other cases, the submodules will require a number of engineers working together. The usual trade-offs between resources and project delivery dates must be considered.

Block design can be broken down into the following five tasks:

- Detailed specification
- Design of the module
- Coding
- Simulation
- Synthesis

Detailed specification involves capturing the design function for the block or blocks in question and tying down the interface definition. Team members should frequently consult each other during this period if there are any ambiguities in interface signal definitions or implemented functions, or if it becomes apparent during this low-level block specification phase that new interface signals or design functions are required.

The design itself involves translating the specification into a design solution. There are several different methodologies for doing this but, in essence, they all involve similar sets of steps. Typically, the blocks can be described by process flowcharts and subsequently defined by block diagrams, timing diagrams and state machine diagrams. There is a temptation to start coding before this initial paperwork design stage is fully worked through. This should be avoided because it can often require the reworking of code to cover the function of the module fully. The design process during this phase is described in detail in Chapter 3, "A Quality Design Approach."

Generation of appropriate design documentation is part of a quality design process. This design documentation should be reviewed before coding has started in earnest. The process of translating the design documentation into code often leads to minor changes in the detailed design approach—perhaps because there were some undetected flaws in the initial approach or perhaps because the designer has thought of a better or simpler way of implementing a particular function. Any design changes made during the coding should be reflected back into the design documentation, and the complete documentation and coding package should then be further reviewed.

It is important to review the code before exhaustive testing is carried out, because there may not be time to rework sections of the code at this later stage.

Simulation and initial trial synthesis should be done in parallel. There is no point to having a fully functioning module that will not synthesize to the correct speed or that translates into an unacceptably large gate count. Encouraging early synthesis also has the advantage that the size of the chip can be monitored from an early stage in the design. If the ASIC size becomes too large, changes to the higher-level architecture or to the design of the module may have to be made. The earlier any such changes are implemented, the smaller the impact will be on the end date.

After the initial synthesis, some initial gate-level simulations should be carried out. This will prove that the RTL (register-transfer-level) code not only functions correctly but that it synthesizes as intended. It also derisks the later top-level gate-level simulation phase.

1.6.2 Other Tasks

The synthesis approach should be documented and reviewed with the ASIC vendor and the synthesis tool application engineers. The documentation should include a generic synthesis script to be used by all the project engineers.

The simulation approach should be defined during this stage of the project and resource allocated to the testbench tasks. The testbench is typically used for both system and subsystem simulations. The development plan should include all testbench work. The testbench should be designed using a flow similar to that used for designing the ASIC modules. The architecture and code should be reviewed. The list of functions that the testbench provides should be documented and reviewed by the team. The plan should allow some time during the system and subsystem simulations for the testbench to be modified or enhanced, because it is difficult to envisage a completely comprehensive set of tests in advance of implementing the detailed design. Ideally, the testbench should be written as synthesizable code, because some tools, such as cycle-based simulators, and other techniques, such as hardware acceleration, are more effective with synthesizable testbenches. As an aid in developing testbenches, a number of testbench generation tools and high-level testbench languages are available from EDA tool vendors (see Chapter 14, "Design Tools").

Any third-party simulation and/or synthesis models required should be identified, and sourcing of these models should be entered like all other tasks into the project plan. Two versions of third-party simulation models may be required—one version that models the functionality to the required maximum level of detail and a second, simpler model.

Simpler models simulate faster and are often adequate for the majority of simulations. The speed difference is not usually particularly noticeable in individual module or subsystem simulations, but the effect can be significant when simulating at the system level. Conversely, detailed models simulate more slowly. This is due to the level of detail being modeled and checked. However, detailed models are required for at least a subset of the simulations because these allow for a more rigorous level of testing and parameter checking.

The project leader should explain the design review process to the team early in the design phase and set out expectations for these reviews. This involves providing guidelines on the level of design documentation required and drawing attention to any design procedures and coding styles being adopted. For companies employing a structured design approach, this type of information is often available in company standards documentation. However, it is usually worth reminding the design team of the most important aspects of these standards and explaining the reason for them. It is also important to explain the reasons for the design review—it is an essential part of a quality design process and not a performance review of the designer.

Full team meetings should start at this stage. They should be held at regular intervals—ideally once per week—at a time that is convenient for all team members. A good time to do this is before lunch because this tends to limit the length of the meetings, and the participants are still fresh. The purpose of the meetings is to ascertain progress against the plan and to keep the team informed of general developments in the projects that are of interest to them. During the team meetings, open communications between team members should be encouraged. It is a good opportunity for team members to highlight any issues or problems that have arisen or are foreseeable.

During this phase of the project, the project leader should start planning the testing of the silicon. This initial planning mainly involves estimating the laboratory space and identifying the equipment that will be needed. These resources should be booked or purchased, if not available.

A further activity, that arises at this stage of the project is that of dealing with the ASIC vendor. The following list identifies some important items that need to be considered at this point and tracked from here on in:

1) Device pin list: The pin list has to be generated, reviewed and frozen many weeks before the final netlist is delivered. The pin list should be agreed by the ASIC vendor, the manufacturing team and the printed circuit board (PCB) design engineers.

2) Package: If the package is new to production, space models (chips with the correct package but with basic test silicon rather than the real silicon) can be obtained from the vendor so that trial PCBs can be manufactured. The quality of the solder joints can be analyzed and the production techniques refined, if necessary.

3) Sample and preproduction volumes: The vendor normally dictates the number of initial samples available. There are a number of different types of samples available with different turnaround and delivery times. For initial testing, it is very important that a sufficient number of development units are available to verify the silicon and the system quickly. With good negotiating, the preproduction volumes can be increased, which can be useful to improve production ramp-up times.

1.7 Subsystem Simulation

Outputs from This Phase:

- Successful run of first subsystem simulation
- Reviewed subsystem simulation specification
- Subsystem module testing complete

Tasks:

- Write and review test list document.
- Write test "pseudocode" (i.e., CPU register accesses, testbench configuration).
- Run simulations.

Project Manager-Specific Tasks:

- Closely monitor plan; arrange regular, quick meetings to discuss progress.

Risks:

- Poor communication of simulation issues between team members can increase timescales unnecessarily.

Subsystem simulation is that part of the project where a collection of separately designed but logically related modules are now connected together and tested as subsystems. For certain designs, subsystem simulation may not be appropriate or necessary. However, for larger designs, it often pays dividends. This phase will typically run in parallel with the module design phase. The subsystem simulation can be carried out by a team containing some of the design engineers who designed the modules in question with the assistance of the testbench engineer. This section is treated under two separate headings: First we look at the tasks and activities of the subsystem simulation team, then we look at the project leader's responsibilities during this phase.

1.7.1 Tasks for the Subsystem Simulation Team

The individual unit modules have already been simulated in the previous module design phase. The reason for an intermediate subsystem simulation phase, rather than just jumping straight from unit to system simulation, is threefold:

First, the subsystem simulations should run significantly faster than the system simulations, so more testing can be done in a given amount of time.

Second, the interfaces to the submodule might allow easier testing of infrequent events. For example, in a communications chip, we may need to process several thousand consecutive data packets in subsystem X to trigger a particular error condition, such as a FIFO overflow. We are interested in establishing that a downstream subsystem, for example, subsystem Y, can detect and manage this error condition gracefully. It may take several hours to simulate several thousand packets to get to this state. In a simulation of subsystem Y on its own, testing the ability of subsystem Y to detect and manage this error condition properly may be as simple as asserting a testbench input representing the error condition from block X. This is all we are interested in from the point of view of testing block Y—we don't want to have to send in several thousand packets through block X each time to achieve this.

Third, the system simulations might require that a particular subsystem is operating correctly before another module or subsystem can be tested. By planning subsystem simulations carefully, tested subsystems will be available when needed for the system simulations.

The subsystem testbench and simulation strategy should be carefully designed and planned. Ideally, the tests will be a subset of the system simulation tests, and many of the same configuration, testbench command and input files can be used during both subsystem and system simulations. This will save time and reduce risks of slips against the plan.

1.7.2 Project Leader Tasks

When the subsystem simulations start, there is often a number of small issues to be resolved to get the first simulations working. The project manager should hold a short meeting every day, so that the issues can be discussed and priorities set. It is very easy for small problems to lie unresolved. Identifying, prioritizing and tracking issues and problems on a daily basis will lead to faster progress and minimize bottleneck hold-ups where a significant number of the team members can be held up by the same problem.

The subsystem simulation phase is the first point where the testbench engineer's work is being used and relied upon. From working in a relatively independent capacity, these engineers suddenly find themselves the focus of attention. They are expected to act in a "help-desk" type role to get the first simulations running and are expected to sort out bugs in the testbench quickly. Any such bugs can hold up several team members at the same time. This is a stage where the testbench engineer is often under considerable pressure from the rest of the team. Given that development of the testbench is frequently assigned to more junior members of the design team who are likely to be less familiar with and less equipped to deal with such pressure, the project leader should ensure that the pressure does not become unreasonable. The daily progress meetings at this stage are a good way of monitoring this and keeping it under control.

The ASIC development plan should be updated at the start of the subsystem simulation phase with a list of the major test areas. The plan should have a milestone of "first subsystem simulation working." This is a significant achievement. It always takes a few days to get the first test working. Tracking each test closely provides a good measure of progress and regular (possibly daily) subsystem test meetings tends to encourage the team to resolve the initial problems encountered quickly.

In parallel with the subsystem testing, work will be carried on combining the entire individual design modules and subsystems together in an integrated full chip-level netlist. For all but the simplest designs, this initial netlist will almost inevitably require rework following simulation. However, it can and should be used in its initial form to carry out trial layouts.

1.8 System Simulation/Synthesis

Outputs from This Phase:

- First system simulation runs successfully
- Reviewed system simulation specification available
- All RTL and gate-level simulations completed

- Initial synthesized netlist available (trial netlist)

Tasks:

- Write and review test list document.
- Write test "pseudocode" (i.e., CPU register accesses, testbench configuration).
- Run RTL and gate-level simulations.
- Record and track resolution of bugs using a fault-reporting system.
- Check chip design rules.
- Write chip device user guide.
- Create synthesis scripts and first trial netlist.
- Create layout floor plan and documentation.

Project Manager-Specific Tasks:

- Closely monitor plan; arrange regular, quick meetings to discuss progress.
- Arrange layout meeting with ASIC vendor.

Risks:

- Poor communication between engineers can significantly increase the time required to get first system simulation running.

The two main tasks during this phase are the top-level simulations and the synthesis of the netlist. As with some of the earlier phases, it is useful to examine separately the activities of the team and those of the project leader. Many tasks during this phase will overlap with the layout and the backend phase that is covered in the next subsection.

1.8.1 Tasks for the System Simulation Team

A number of tasks must be completed before system simulation can start. A prerequisite of the top-level simulations is a full top-level netlist that includes the I/O pads. The top-level netlist should be carefully reviewed prior to the start of the simulations to eliminate simple connection errors. The synthesis team typically generates this netlist.

The top-level testbench is a further prerequisite for system simulations. The testbench engineer should provide a basic user manual for the testbench, detailing such things as how to set up testbench configuration files and how to set up different operating modes for complex transactors. The testbench engineer should also give a formal presentation to the team before simulation begins, explaining the architecture of the testbench and how best to use it.

It is usually a good idea to split the team into a number of subteams of one to three people each for each of the areas in system simulation. Each team is assigned a different area to focus on during the system simulations. Each team must define a list of tests that comprehensively test its area. The list of tests should be reviewed by the full team to ensure that there are no obvious omissions. The test list will typically include a subset of the tests carried out during unit and subsystem simulations as well as any additional tests that can be run only at the chip level. For some modules, the time taken to create truly representative input stimuli at the module level can be prohibitively long. For these modules, chip level testing is the only way to validate the module fully.

The system simulations should be run first on the RTL code and subsequently on the gate-level netlist, when the gate-level netlist is available from the synthesis team. The RTL simulations should be carried out first, because these simulate faster and the RTL description is always available before the gate-level netlist. The RTL simulations prove that the entire chip implements the functions and algorithms defined in the top-level specification. Gate-level simulations are discussed in the layout and backend phase that is covered in the following subsection.

For large designs, where it is difficult to plan the coincident completion of all blocks, initial top-level system simulations can be done without all the modules being available. In this case, simple dummy modules to allow the netlist to elaborate or compile can replace any missing modules.

The project should develop some guidelines to ensure a systematic and consistent approach to system simulation. These can be drawn from the best practices and experiences on previous projects and should be improved and updated with each passing project. Among other things, the guidelines should cover test directory structure, test naming conventions, netlist version management, result logging, etc. One method of approaching directory structures is to have separate directories for each test category. These can be further divided into subdirectories, as appropriate. Ideally, the test file names will match the test names in the test list to allow easy cross referencing. For example, a test file named *error_corrector_4.3* would correspond to test 4.3 in the test list. Clearly from its title, it involves testing some aspect of the error corrector.

It is important that tests are run on the latest version of the netlist. There are, of course, limits to this. If the netlist is being updated on a daily basis, it may not be practical or efficient to rerun all tests on all new versions of the netlist. In such a scenario, perhaps releasing the codebase once a week, for example, is more appropriate. The

RTL and gate-level code should be under official revision control, and the simulation teams need to understand the simulation directory structure and to know how to link to relevant versions of the netlist. During the system simulations, changes to the VHDL or Verilog code should be done in a controlled way so that everyone knows when changes have been made and what tests they are likely to affect. Time is often lost trying to understand why a particular module under test has stopped working when the problem has been caused by changes to other modules. The team should agree when and how code is released, and who is allowed to release it. These issues are discussed further in Chapter 7, "Quality Framework."

Officially tracking problems and their resolution is essential to ensure that problems are fixed before the final netlist is sent for fabrication. The project manager can also use tracking statistics as an indication of how well the netlist has been tested. At the start of system simulations, the rate of problems identified will be high. Toward the end, the rate should tend toward zero. Clearly, the netlist should not be released if the rate of problem detection is still rising. Last-minute changes to the netlist may not affect all modules. In such cases, it is valid to rerun only a subset of system simulations to reduce timescales.

There is a number of steps that can be taken to reduce the duration of the system simulation phase. Create self-checking scripts that will automatically run a series of system simulations and log results. This is useful for rerunning the tests following a change to the netlist. Because the tests are invoked via scripts, it is easy to set them off overnight, for example, instead of having to run each in series interactively. Wherever possible, use the same tests for RTL and gate-level simulations. Limit the use of internal probing and node forcing. Nodes that are available in the RTL code may not be available in the gate-level netlist. Start trial system simulations early, with the testbench engineer and a single design engineer. This ensures that time taken to iron out the basic testbench problems does not affect the entire system simulation team. It is a form of testing the testbench and gives an early indication of how user-friendly the testbench is and whether the documentation on using the testbench is adequate.

Typically, the system simulations will take several man-months of effort to complete. For complex designs, the system simulations can never fully test the netlist without incurring very long simulation times. In these cases, an early sign-off can significantly reduce the time before production silicon is available. The idea here is to fabricate the netlist before the all system simulations have been completed. However, a subset of both RTL and gate-level simulations must have been completed before the early sign-off takes place. This early silicon is used to test the functionality of the design rather than testing at all environmental conditions. Therefore, the netlist can be sent for prototyping before the gate-level netlist has achieved the required synthesis speed. The advantage of this approach is that, once the trial silicon is available, testing with test equipment is much faster than testing by simulation. Consequently, a very large number of tests can be executed in a fraction of the equivalent simulation time. There should be a planned second version of the silicon that is fabricated only after initial testing of the first silicon has been completed. Avoid the tendency to fabricate the second silicon too early before the initial testing has been completed, because bugs will often be found toward the end of the testing.

Toward the end of the top-level system simulations, the list of tests and the results should be carefully reviewed, because the quality of these tests have a direct bearing on the success of the project. Often, as the test teams become more familiar with the complete design, they will come up with additional tests that were not thought of at the time of formulating the initial test list. If these new tests are testing functionality that was not otherwise being tested, they need to be added to the test list.

1.8.2 Tasks for the Synthesis Team

Synthesis is the process of translating the RTL description into a gate-level netlist using a synthesis tool. The synthesis team has the job of integrating the top-level modules together into a top-level RTL description, then generating the gate-level netlist from this. The top-level RTL should be fully reviewed, paying particular attention to the pads. The definition of the power pins is particularly important because they cannot be fully simulated. Synthesis scripts are used to guide the synthesis tool and drive the synthesis process. The scripts provide information to the tool on physical and timing characteristics of the circuit and its signals. This information is essential to enable the tool to map the RTL code properly to a gate-level netlist that meets the required timing characteristics. The scripts contain information on clock signals (period, phase, duty cycle, aspect ratio, relative phase, etc.), input signal driving strengths, output signal loads, input and output timing requirements, multicycle and false paths, etc. The top-level synthesis scripts can be relatively complex and take a significant amount of time to perfect. Fine-tuning and tweaking of input parameters and constraints is usually required to get the desired end result. Synthesis and synthesis scripts are covered in Chapter 6, "Synthesis." The scripts should be reviewed internally by members of the design team and in-house synthesis experts. They should also be reviewed externally by the ASIC vendor and the synthesis tool application engineers.

Test insertion and test vector generation are usually considered part of the synthesis process. Test vectors can take a significant amount of time to generate and simulate. The test vectors must be run on the gate-level netlist using best-, typical-, and worst-case timings and can take many days to simulate, depending on the size of the netlist and the processing platform. Test insertion should be done as soon as the top-level netlist is available. This allows the maximum amount of time to get the tools working correctly. After test vector generation, the fault coverage can be assessed. Doing this early in the system simulation phase allows time for possible changes to the code to increase fault coverage. Functional vectors and parametric tests must be planned into the synthesis phase. These tests can be tricky to generate and can cause delays to the ASIC sign-off, if left to the last moment. Some ASIC vendors require extensions to the basic Verilog simulator to generate the test vectors in the required format. This requirement should be identified early and the extension incorporated and tested in the design environment.

The layout flow should be discussed and agreed on between the team and the ASIC vendor early in the project. Trial netlists should be sent to the vendor at this stage and the results analyzed (see the following subsection on postlayout simulations). The layout will be improved by ensuring good communications between the ASIC team and the ASIC vendor. The design engineers and the layout team should jointly develop a floor plan, and layout information should be generated and documented. Relevant information includes the clocking strategy, identification of blocks requiring specific placement on the die (typically for speed or noise reasons) and identification of critical paths. Timing-driven layout is a useful method of increasing the timing performance of the final silicon by focusing on the most time-critical paths first. Some synthesis tools are capable of automatically generating constraint files that can be read by the timing-driven layout tools. For deep submicron designs, the synthesis and layout tasks constitute a significant area of risk in the project. This is because the technology geometries are so small that interconnect delays become the dominant part of timing delays, rather than the gate delays. This can result in large differences between prelayout and postlayout timing figures, consequently making it difficult to achieve timing closure on the design. New tools that take floor-planning information into account as part of the synthesis operation are starting to appear on the market to address this particular problem. For large deep submicron designs, it is essential to agree early on with the ASIC vendor how the timing closure will be managed. It requires closer cooperation with the ASIC vendor than was traditionally the case in previous generation designs and requires careful planning of the division of roles and responsibilities between the ASIC vendor and the design team.

1.8.3 Project Leader Tasks

The ASIC team leader faces the same sort of issues during the system simulation phase as were faced during the subsystem simulation phase (see above). Getting the first system simulation running properly is a major milestone. It is advisable to hold daily progress meetings from the start of this phase to identify and track problems. Common problems include testbench bugs or functionality limitations, inadequate testbench documentation, intermodule interface bugs, lack of hardware resources (hard disk storage and workstation memory), shortage of available simulation licenses, etc. These problems need to be prioritized and the task of resolving each problem assigned to various team members.

The management of the ASIC vendor becomes critical during this stage. This is covered in the following section, "Layout and the Backend Phase." Understanding the layout flow can help to reduce timescales. The project manager should arrange meetings to agree on the layout flow and discuss layout issues, synthesis scripts and delivery dates. It is useful to have an understanding of the task breakdown in the vendor's layout group and how it is addressing these tasks.

1.9 Layout and the Backend Phase

This section documents those tasks that are the responsibility of the ASIC vendor.

Outputs from the ASIC Vendor:

- Postlayout timing/capacitance information
- Information for silicon manufacturing

Tasks for the ASIC Vendor:

- Layout of the trial and final netlists
- Checking the netlist and test vectors for errors
- Generation of postlayout timing/capacitance information

This phase deals with the layout work that is carried out by the ASIC vendor. It also covers the work that the design team must do to support the layout team and the tasks that must be done following the layout. This subject is described in more detail in Chapter 10, "Dealing with the ASIC Vendor." This phase overlaps many other phases because layout should start as soon as an initial top-level netlist is available. Indeed, some critical modules may need layout before the top-level netlist is available.

The ASIC vendor takes the VHDL or Verilog netlist and converts it into a physical layout. The process involves a number of complex tools, and the level of risk increases with the size and speed of the design. The risks are further increased if the ASIC is substantially different from ones that the vendor has previously processed. Differences could include different technology, very large gate counts, or the use of complex compiled cells, etc.

Trial layouts will reduce the risks by identifying any potential problems as early as possible. The project should plan for at least two trial netlists, and the number should be agreed on with the ASIC vendor at the start of the project. It is important that the trial netlists are passed through the complete layout flow.

The first trial layout should identify any major problems, such as design rule errors, modules that cannot be routed and major postlayout timing violations. The first trial netlist does not need to meet all final timing requirements. However, the extent to which timing requirements are violated at this stage will give an idea of how realizable the design is. It is also useful to examine the extent to which prelayout and postlayout timings differ.

The second trial netlist should meet all timing requirements. Meanwhile, some initial system simulations should have been run on the RTL and gate-level code. If the project is lucky, this second trial netlist becomes the final netlist, and the sign-off can be completed earlier than planned. If simple bugs are found during the system simulations, these can often be fixed in the layout, using an engineering change order (ECO). These ECOs define gate-level changes to the netlist, which can be done manually by the layout engineer. The ECO approach varies for each ASIC vendor, and the options available for ECOs should be discussed at the start of this phase.

The project team and the ASIC vendor should agree on a release naming convention for the VHDL or Verilog code. It is important that the ASIC vendor has enough time to complete the layout flow before the next netlist is sent, so the number of revisions should be kept to a minimum, even if the netlist seems to be changing on a daily basis.

The layout process involves floor planning, cell placement, clock-tree insertion, routing and timing analysis. Postlayout timing information is generated as part of the layout process, and this is used for postlayout simulation and synthesis.

As mentioned in previous sections, timing delays can be significantly longer after layout, especially for deep submicron technology. It may take several iterations of design, synthesis and layout to meet the target timing requirements. Newer synthesis tools can generate timing and routing constraints that are used to guide the placement and routing layout tools.

1.10 Postlayout Simulation/Synthesis

Outputs from This Phase:

- Final netlist sent for layout
- Test vectors (IDDQ, scan and functional) tested and sent
- Postlayout simulations and static timing completed
- Chip sign-off

Tasks:

- Synthesis, test insertion and test vector generation
- Generation of a layout document
- Support layout (floor planning, checking timing, etc.)
- Resynthesis after layout (fixing overloads, timings violations)
- Running gate-level simulations and static timing analysis with final netlist, using back-annotated timings

Project Manager-Specific Tasks:

- Arrange meetings with layout engineers/synthesis engineers.
- Review progress/milestones of layout.

Risks:

- Pin-out errors are common—review this several times.
- There may be issues with layout (routing, timing after layout, etc.)—do trial netlist as early as possible.
- Test vector generation can take a long time—start generation of vectors early.
- Gate-level simulations are polluted with unknowns (Xs)—emphasize the need for reset conditions on all registers early in the design.

During layout, a number of files are generated that can be used to analyze the impact of the layout on the circuit. The files are back-annotated onto the netlist, and some processes are rerun. There is a number of steps in the layout process. Some of these files can be extracted after the initial layout stages of floor planning and cell placement. This gives an indication of the layout but is not yet fully accurate, because it does not contain the routing information. After the layout has been completed, a custom wire-load file, a capacitance load file and a standard delay format (SDF) file are generated. Normally, an updated netlist is also sent back by the vendor, because the clock tree will have been added during the layout.

The custom wire-load gives the average wire length for a module or chip. Before a custom wire-load is available, the synthesis tool uses an estimated wire-load file. The wire-load values in this are estimated based on the synthesized gate count. The custom wire-load file from a trial placement can be used as the starting point for a complete resynthesis, if desired. Alternatively, the custom wire-load from a final or almost final layout can be used in combination with an SDF file for minor resynthesis or in-place optimization. In this scenario, the SDF timing information is used for all cells and nets that do not change, whereas the wire-load file is used in calculating delays for the small number of new nets created in the resynthesis.

The SDF file defines the time delays between each pin in the netlist. This includes delays through the cells and interconnect delays. It is used for both synthesis and simulation. The synthesis tool will use it to do static timing analysis and for resynthesis to solve timing violations. At this point, the synthesis script should now use a propagated clock, rather than the ideal clock that was used prelayout.

Postlayout timing analysis is a critical stage in the project and should be carried out as quickly as possible and the results analyzed. This is typically done using a static timing analysis tool. Results from the trial netlist could indicate the need for a different layout, changes to the code, or resynthesis. With current small geometry technologies, the interconnect delay is increasingly important in defining the critical timing path in large gate-count designs. The static timing analysis will highlight particularly long nets. Sometimes, altering the floor plan or cell placement, or simply rerouting some nets can resolve timing violations. Another approach to resolving timing problems is to resynthesize with back-annotated SDF, custom wire-load and capacitance files. The synthesis tool will try to meet the timing requirements based on the actual delay and capacitance file.

Anything other than trivial changes to the code to fix timing problems will inevitably require changes to the plan and possibly also changes to the test list. The test plan can be rescheduled to focus initial tests on parts of the design that will not change. This minimizes the effect of any redesign on the testing schedule.

The capacitance file defines the capacitance that each cell output drives. The technology library will define a maximum capacitance that the outputs can drive. If this is exceeded, the ASIC vendor's design rule check (DRC) will fail. The synthesis tool can read the capacitance file and rebuffer the overloaded nets. It can be set to allow rebuffering only (preferred, because it has least impact on the layout but might not fix all violations) or to allow restructuring so that an overloaded net is replaced by a number of parallel nets. The overloaded nets should be fixed before timing analysis is done, because the overloaded nets result in poor timings. Traditionally, the project team, rather than the ASIC vendor, fixes design rule violations, but some ASIC vendors have tools that can automatically fix design rule violations that have arisen as a result of layout.

Tip: The cell timing models are based on standard industry-defined worst-case temperature and voltage. If the design team is confident in advance that the power consumption or operating voltages are, in reality, going to be more favorable than the standard worst-case conditions, custom worst-case operating conditions can be used for the cell timing models. This helps to achieve the required timing performance. If the cell models are not designed to take variable operating environment parameters, the ASIC vendor will need to provide a custom synthesis library. Even if the cell models can take variable operating environment parameters, it is important to establish with the ASIC vendor that the models are sufficiently accurate at these nonstandard operating points.

1.11 ASIC Sign-Off

Outputs from This Phase:

- Chip sign-off

Project Manager-Specific Tasks:

- Check sign-off documentation.
- Get agreements from different departments for silicon quantities.

After postlayout simulations and synthesis have been completed, the netlist is sent for fabrication. This is referred to as *ASIC sign-off* because there is typically a document that is signed by the team and the ASIC vendor. The document clearly states the netlist revision ID, the required test vector files, agreed sample quantities and commercial aspects. This document should be studied several weeks before the sign-off date so that all items can be completed on time. Representatives from a number of departments, including development, manufacturing and marketing, must complete the document.

Before sign-off, the ASIC vendor must run checks on the netlist, layout and test vectors. The vendor will normally require that the test vectors be simulated before sign-off. This can be a lengthy process. With some persuasion, the vendor may agree to send the netlist for fabrication before these simulations have completed. However, because this prevents the vendor from properly testing the fabricated silicon, the vendor will disown responsibility for fabrication problems until the test vector simulations have all passed.

1.12 Preparation for Testing Silicon

Outputs from This Phase:

- Reviewed evaluation plan
- Test equipment commissioned before silicon available
- All tests prepared (hardware, software and automation)

Tasks:

- Write and review evaluation test list/plan.
- Write tests.
- Plan and implement test automation.
- Reserve test equipment.
- Design or outsource design of nonstandard test equipment.
- Commission test equipment before silicon is ready.
- Define requirements for evaluating silicon at different voltages and temperatures (environmental testing).
- Define method for recording, analyzing and resolving bugs.

Risks:

- This can be a time-consuming task—allow appropriate planning and start early.
- The specification for the ASIC is not available/inaccurate when planning this phase early in the project.

Preparation for testing of the silicon is, in principle, a straightforward task. However, it is all too easily ignored while focusing on planning the more complex stages of the ASIC development cycle.

The time required to verify the silicon can be significantly reduced if some preparation work is carried out before the device arrives. The preparation usually requires significant effort and should be carefully planned. The work is often done in parallel with top-level simulations, and the risk is that it may be pushed aside as more pressing issues with the simulations arise. To avoid this, the preparation tasks are ideally assigned to engineers who have little involvement with the top-level simulations. The preparation requires the involvement of a number of different engineering disciplines, including ASIC, software, PCB and test engineering.

There is a number of aspects to preparation for testing. Laboratory space must be booked in advance. Standard and specialized test equipment must be booked and purchased or hired, if not available in-house. Special attention should be paid to the purchase of specialized test equipment, because lead times may be long.

Test PCBs must be designed, manufactured and populated in advance if test boards, rather than IC testers, are being used. If the ASIC is to form part of a production PCB, the design of the board should be done in conjunction with the ASIC engineers. The ASIC pin-out should be defined with help from the PCB designers because the pin-out can have a major impact on the PCB track routing. IO timings, output pad load capacitance, output voltage levels and input requirements should be discussed with the PCB designers. The ASIC should contain circuitry to allow easy debugging. This should allow access to internal status registers, access to internal nodes via output test pins, built-in self-test, error-checking circuits, etc. It is important that these debugging features are discussed with the PCB designers so that test equipment can be easily connected to the ASIC. IC clips can be bought that fit over some ASICs, giving access to the chip inputs and outputs. However, for high pin counts, these can be very expensive and not fully reliable.

Modern ASICs can often have packages with very high pin count requirements. There is a range of packages available, such as BGAs, micro BGAs, flip-chip, or TAB. These give cheap, reliable packaging but require some experimenting to achieve high yields in the PCB assembly process. The production team can practice on empty packages before the silicon is available, which ensures that the first samples are mounted on the board as quickly and reliably as possible. It is important that the number of PCBs required is agreed on at an early stage to allow time for the purchase of components and for manufacturing resource to be allocated.

Development of test software must be planned and the code written and tested as well as it can be tested in the absence of the silicon. The testing can be done using a software emulation model of the ASIC/board. Alternatively parts of the test software can be tested within the simulation environment (see Chapter 5, "ASIC Simulation and Testbenches").

Automating the tests will significantly reduce test times but, of course, requires additional effort. The automation can be done using specialized test equipment or based on PCs/workstations with application-specific test software.

The test software is based on the test list specified by the design and test teams. The test list should be reviewed by the team to ensure that it is adequate to test the ASIC comprehensively. One approach to creating the test software is to provide a set of common low-level driver routines for setting up various chip configurations and register settings. If these routines are sufficiently comprehensive and well documented, it is then easy for members of the test and design teams to write a layer of higher-level test code to sit on top of these. For register-intensive designs with many possible operating modes or configurations, it can be useful to design the test code to initialize the chip based on test configuration files. Changing the operating mode can then be as simple as changing a number of ASCII text fields in a configuration file.

SoC or large designs may have IP blocks from third-party vendors. These blocks should be initially tested wherever possible in a stand-alone mode based on a test suite provided by the vendor. Once the block has been tested in a stand-alone mode, it can then be tested operating within the integrated device. The stand-alone tests may require specialist test equipment, and this should be sourced and tested well before silicon is available. Some IP providers produce reference boards that can be used to check the test equipment and test setup.

Sometimes, part of the test list includes interoperability testing. This means testing the silicon within its system environment with a range of existing products. These tests must be defined at an early stage so that the existing equipment can be acquired.

During testing, problems will be found. There is a number of potential sources of such problems, such as PCB errors, software bugs, test equipment failures, etc., in addition to any possible problems in the silicon itself. However, each failing test must be recorded in a database that can be easily accessed by the entire team. The fault-tracking database must be created and the bug tracking process defined. A template for the test report is also needed before the start of the evaluation.

1.13 Testing of Silicon

Outputs from This Phase:

- Fully tested silicon working in a real application
- Test report generated and reviewed

Tasks:

- Run the tests.
- Track test failures in a fault report database.
- Analyze failing tests.
- Identify workarounds for ASIC bugs.
- Identify netlist changes for ASIC bugs.
- Evaluate silicon at different voltages and temperatures (environmental testing).
- Do interoperability testing.

Risks:

- A shortage of working PCBs can significantly increase the time to complete the initial testing of the silicon.
- Modifications to the PCBs, as part of the debugging process, can make them unreliable, resulting in a shortage of working PCBs.

The preparation for the testing of the silicon should result in a test list and project plan for this crucial phase. The silicon will typically need to be mounted onto a printed circuit board. Ideally, a number of printed circuit boards with the new ASIC should be manufactured. It is important that an adequate number of working PCBs are available for the ASIC, software and PCB engineers.

The first tests should prove the basic operation of the silicon. Initial problems will always be encountered when running the tests. However, these may not be due to the design. Faults in the PCB, problems with manufacturing of the silicon, assembly of the board, software bugs or faulty test equipment can all cause tests to fail. To identify the problem area quickly, there should be an agreed-on approach when tests fail. Some options are suggested below:

- Run the same tests on multiple boards and ensure that the failure mechanism is repeatable.
- Check the test setup.
- Check register configurations (read the internal registers).
- Check internal ASIC debug registers.
- Check external ASIC signals and internal debug signals that can be multiplexed to output pins using logic analyzers and oscilloscopes.
- Create a simulation test that mimics the failing test.
- Analyze the VHDL/Verilog code for the failure mechanism.
- Brainstorm ideas for devising different tests that will identify the problem.

- Test at different voltages and/or temperatures to determine whether the problem is related to environmental conditions.

Which of the above options are followed and in what sequence will depend on whether the failure occurs during the initial tests or after many tests have been successfully run. Test failures should be noted in the fault report database.

One of the most important administration tasks is to keep an accurate track of the location and status of the PCBs. Whenever a test is finished, the result and the serial number of the board should be noted in the test report. This allows the test setup to be reproduced at a later date when tracking problems. A spreadsheet should be created which defines the status and location of each PCB, and this should be available on a common server/intranet. It is useful to record which boards have passed/failed which tests on a regular basis. It is also important to record the modification state of the PCBs. This is often the responsibility of the PCB engineers. These engineers should also be responsible for having PCB manufacturing problems fixed.

1.13.1 Project Leader Tasks

The evaluation of the silicon is a complex task because it involves ASIC, PCB designers and software engineers all working together. During the planning of the evaluation, the resources from the different disciplines should be identified and tasks assigned. The ASIC leader should ensure that adequate PCB and software resource is available, especially at the start of the evaluation.

For the first few days of the evaluation, the ASIC leader should check the progress several times a day. A list of issues and problems should be maintained, which should be analyzed daily to assign priorities and resources. After a few days, progress can be tracked daily. If a major bug has been identified within the silicon, the ASIC leader should arrange a meeting to brainstorm ideas for solving the issue without the need for changes to the ASIC netlist. It is important that PCB designers and software engineers are present at the meeting because they may be able to solve bugs by changes to the PCB or software. These types of changes allow the product to be launched earlier than if an ASIC respin is needed. However, they can often have some impact on the performance of product features or product cost. The ideas and consequences should be presented to marketing engineers and senior managers. Changes to the ASIC netlist should also be identified, and the ASIC leader should estimate the time before new silicon would be available.

The project leader should ensure that each engineer is completing the test reports and filing problem reports into the database. These can be somewhat tedious tasks, but are necessary for the successful completion of the product. The list of open bug reports should be reviewed periodically with senior management. Two ratios are useful as a guide to the level of testing and the confidence in the silicon. The first is the number of completed tests, compared with the total number of tests. The second is the rate of finding bugs, compared with resolving bugs.

On completion of the silicon testing, the test results should be reviewed by the team.

1.14 Summary

With more and more of the electronic circuitry in new products integrated into ASICs, the ASIC design activity in a new product development can appear a daunting task to senior managers, project team leaders and ASIC specialists alike—the latter often being expert in one or more parts of the ASIC development process but lacking an adequate understanding of other parts. However, like most engineering problems, the way to approach this one is to break it into a series of smaller, more manageable tasks or phases. Each phase consists of a series of related design and support activities that can be collectively grouped and classified under one heading.

This first chapter attempts to capture the entire ASIC development process in overview form by dividing it into a number of mainly sequential phases and describing the main aspects of each such phase. It is a useful chapter for the entire project team to read before embarking on a new ASIC development. More detailed information on specific phases is provided in the subsequent chapters of the book.

## CHAPTER 2

Design Reuse and System-on-a-Chip Designs

2.1 Introduction

To meet the demand for increasingly large ASICs with increasingly aggressive development timescales, new ASIC developments will typically have to reuse modules that were designed in previous projects. Large system-on-a-chip (SoC) designs additionally usually necessitate reuse of external third-party IP cores. Module reuse is, therefore, becoming an ever more important feature of ASIC design methodologies.

Design reuse is not a new concept. In fact, the use of standard cell libraries is a long-standing application of design reuse. However, what is a more recent development is the concept of reusing larger and larger modules as building blocks to speed up the development of large system chips.

Design reuse presents many challenges. There are always trade-offs. The fine tuning or optimization required to achieve a specific performance goal is often at odds with the requirement of keeping the design generic enough to be reusable in other applications.

In addition to technical challenges, there are also cultural challenges. Design engineers are sometimes less than enthusiastic about design reuse. This is partly because they perceive themselves as creative and like to do original work. It is up to management teams and project leaders to explain the need for and benefit of design reuse and to create a reward system that encourages a design reuse culture.

Good documentation is fundamental to design reuse, as is designing with reuse in mind from the outset. The first part of this chapter describes reuse documentation, and the second part provides some practical tips and guidelines for successful reuse. The final section considers some of the issues surrounding reuse of external third-party IP in SoC designs.

2.2 Reuse Documentation

In the absence of good design documentation, reuse is not only risky, but also an unwelcome chore for the designer who is reusing the module. When a module is properly coded and documented, designers tend to be considerably more open to the concept of reuse. Therefore, an appropriate level of reuse documentation is not only an essential part of a design reuse methodology—without which reuse is difficult or even impossible—but is also essential to maintain designer morale in an environment where design reuse is the best strategy.

Ideally, it should be possible to reuse a module without the need to understand the implementation in great detail. The functions and limitations, however, must be fully understood. The reuse documentation should, therefore, provide detailed, concise information on the module interfaces and sufficient description of the internals to enable minor modifications to be made, if necessary, for specific reuse applications.

In summary, the design reuse documentation should resemble third-party data sheets provided by manufacturers for their devices. However, in contrast to such data sheets, the reuse documentation must also provide sufficient details of internals to allow minor modifications, where necessary.

The reuse documentation should contain the following components:

2.2.1 Functional Overview

This section should provide an overview of the function of the module, e.g., "this module collects statistics on traffic levels observed at each of the network ports and stores these in a number of counters each representing different categories of traffic," or "this module decodes the incoming data stream and extracts framing and synchronization signals." The functional overview should not go into design detail. It is simply a short description of what the module does.

2.2.2 Interface Description

This should contain a description of the interface signals, typically in tabular format. Each signal should be identified as input, output or bidirectional. It is also useful to provide an additional column of information on each signal, giving details of the signal's characteristics, for example, "single cycle pulse" or "must be asserted no later than

N cycles after signal *X* is deasserted." A maximum load should be specified for output signals.

2.2.3 Implementation Details

This section should include block diagrams and implementation notes. In simpler cases, the code itself may be self-explanatory, but block diagrams and implementation notes are usually necessary. The level of detail in an implementation overview should not be pitched too low. If needed, lower-level detail can be figured out by analyzing the code itself, reading comments within the code, or examining any additional accompanying documentation, such as flowcharts and timing diagrams. However, any unusual tricks or techniques used in the design should be referred to in the implementation overview. Although an aim of reuse is the ability to reuse the module without modification, this is not always realistic. It is, therefore, important that the implementation overview, combined with the other components of design reuse documentation, is adequate to facilitate tuning of the design, where necessary.

2.2.4 Timing Diagrams

If a module is in any way complex or is difficult to describe in text-only form, timing diagrams are often the best method of describing parts of a design. As a minimum, it is usually necessary to provide timing diagrams for some of the interface signals. In certain cases, it is useful to provide internal timing diagrams also.

Often, a pipelined design approach is used to increase the data throughput of a module. The timing diagrams should show the maximum data throughput and the latency of the module.

2.2.5 Test Methodology

The module documentation should include a section on test methodology (for example, full scan, partial scan or functional test vectors). It should also identify any special features used to improve partial scan or functional test vector approaches (for example, splitting large counters into smaller parts). Any built-in self-test (BIST) logic should also be noted. The documentation should describe how to start the BIST and what registers are available to determine that the test has completed and has passed or failed.

2.2.6 Clocking Strategy

A full description of each of the clocks in the modules and their interrelationships (for example, whether they are phase related or asynchronous) should be provided. The test methodology for handling different clock domains should also be described.

2.2.7 Source Code Control Strategy

A good design flow will include some form of source code control. The reuse documentation should describe the type of source code control system used and should indicate where the code is archived. In addition to source code itself, the synthesis scripts and general design documentation should be maintained in a document control system.

2.2.8 Synthesis/Layout Approach

The documentation should include synthesis and layout notes. Ideally, the synthesis constraints file will contain all of the synthesis requirements. However, where necessary, additional documentation may be required to explain specific requirements that are not easily captured in a constraints file or perhaps to explain complex constraints that cannot be adequately described by short comments in the synthesis scripts.

The synthesis scripts should be as standard as possible. However, if module-specific actions are required, for example, the need to flatten sections of the design or the setting of multicycle paths, these should be highlighted in module-specific parts of the script. Alternatively, they can be read in from separate module-specific files.

2.2.9 Module Validation Approach

The documentation should list a set of tests that fully test the module. A short description of what each test checks and, optionally, how it checks it should be included in this section. Because simulation time is often significant on large chips, it is common that only a subset of the complete behavioral-level test suite is rerun at gate level. The test list should identify this subset.

2.2.10 Programmer's Reference Guide

When the module can be configured or controlled by a microprocessor, a programmer's reference guide must be part of the reuse documentation. The guide should define the function of each configuration register bit. Where appropriate, it should also include application notes that specify which combinations of register settings are required to set up different operating modes. Ideally, the documentation should be accompanied by software driver code that allows easy programming of operating modes through a programming interface. The software also needs to be documented.

2.3 Tips and Guidelines for Reuse

With the increasing interest in reuse methodologies, there has been a significant growth in the amount of published material that is available on the subject. In addition to specific textbooks, the topic is receiving increasing coverage in engineering magazines and journals. A simple Web search will reveal a surprisingly large number of references. However, reuse is not an exact science, and there is no one correct way to approach it. Most available material could be categorized under the heading of tips and guidelines, and this is what the remainder of this chapter addresses. How applicable such guidelines are depends on company design methodologies and specific design details.

One fundamental guideline is that it is important to design the module for reuse from the outset. In fact, the process ideally begins at the architecture stage. It is much more difficult to adapt a module for reuse that was not designed for reuse from the start. Fortunately, many of the tips and guidelines that apply to design reuse are simply good design practice in the first instance, whether or not design reuse is being considered.

2.3.1 Company Standards

It is important that reuse methodology is consistent throughout the company. The creation of intranet Web pages dedicated to reuse information is a method of getting a design reuse culture up and running. The Web pages can contain reuse standards, rules and guidelines, along with a library or catalog of modules available for reuse.

2.3.2 Coding Style

The design should be coded using a standard coding style. Designers familiar with this coding style will find it easier to recognize specific constructs and understand the underlying hardware structures. The code should be adequately commented, bearing in mind that the eventual reuser of this code might not be the original designer. A side benefit of using a standard coding style and good use of embedded commenting is that reviewing the design—also an important aspect of reuse itself—becomes easier.

Designers should avoid writing code that is unnecessarily complex. The temptation to write overly clever compact code should always be countered by consideration for the reuser of such code—will it be easy to understand? On a related but more specific note, in VHDL implementations, designers should be judicious about the number of types and subtypes they create. Too many new types can be confusing for the reuser and can make the code appear unnecessarily complex. If new types are created, it is helpful if the type name is descriptive of the nature of the type.

2.3.3 Generics and Constants

The VHDL coding language allows the use of generics. This allows modules or functions to be designed with variable input parameters that are set according to the needs of the particular context in which the module or function is used. For example, rather than design a byte-wide FIFO that is 16 words deep, a designer can design an N-bit wide and M-word deep FIFO. This same FIFO can then be used in multiple different configurations throughout the design by simply defining the generics N and M in each instantiation. Similarly, the same FIFO can be used in future designs that do not use the same bus width or that require a different FIFO depth.

Wherever possible, generics should be used to make a module as flexible as possible. However, some synthesis tools cannot handle certain types of generics, so the approach needs to be tested early on with the intended synthesis tools before being adopted for all modules.

A similar concept to the use of generics is the use of constants. Design modules often use predefined constants, such as register address values, coefficients in mathematical calculations, etc. These constants should be defined in a dedicated file where they can be associated with user-friendly names that are used in the source code to represent the constant values when they are needed. During compilation, the constant definitions file is used to resolve references to the constants as they occur in the source code.

For example, instead of specifying an input bus width of say [15:0] on an input to a module, why not specify an input bus width of [BUSWIDTH-1:0], where BUSWIDTH is a constant defined in a central constants file? If any processing of this bus is carried out in the module, it may be possible to keep this in the generic constant format. This allows the module to be readapted for slightly different bus widths.

Another common situation where use of constants can improve the design's reusability is in address decoding applications. For example, in programmable ASICs, CPU-accessible registers are accessed by decoding an address or address range that is specific to the register in question. The address decoding logic should use user-defined constants to represent the addresses, rather than putting in hexadecimal values. The mapping of constants to register hexadecimal addresses can then be contained in a separate constants definition file. In this way, when a module containing registers is reused in a chip with a different address map, the address changes can be made in one central mapping file. An added benefit of using constants is that well-chosen constant names will make the code easier to understand than their corresponding hexadecimal values. For example, a piece of code that reads:

```
IF ( address = display_ctrl_addr ) THEN ....
```

is more informative than a piece of code that reads:

```
IF ( address = $FFCB )  THEN ....
```

2.3.4 Clock Domains, Synchronous Design, Registers and Latches

This heading is worthy of a chapter in itself, if not an entire book. However, it is still worth highlighting some of the issues that the heading covers.

The design should be made as synchronous as possible. Asynchronous interfaces are more difficult to synthesize and may be misunderstood more easily. Where asynchronous interfaces between two clock domains are necessary, it is advisable to try to group these into separate modules. This way, they can be easily identified in synthesis and given any special treatment they require. By isolating them in separate modules, it is also easier for the design team to review them.

In addition to limiting the number of clock domains to as few as possible, it is preferable to avoid using gated clocks in the design. Gated clocks always present synthesis complications and require more attention—and, consequently, time—in the synthesis process. Therefore, where possible, they should be avoided. However, if power consumption is an issue, there are advantages in using gated clocks. In certain designs, therefore, a power versus ease-of-synthesis trade-off must be made.

Registers should be used in preference to latches, where possible. In older designs, when gate count was critical and design speed was dominated by gate delays, rather than interconnect delays, latches were sometimes used in preference to registers. However, as a result of shrinking geometries, gate count is typically not as critical, and design speed can be influenced as much by interconnect delays as by gate delays. Registers are simpler to synthesize and are less ambiguous to synthesis tools than are latches. There are, of course, instances where latches provide specific functionality that a register cannot, and in such cases, a latch may be the appropriate choice. The guideline is to use registers in preference to latches, where the choice exists. From a synthesis point of view, if a latch must be used, it is important to consider the timing path through the latch to the next synthesis element endpoint when the latch is in transparent mode in addition to considering the setup timing path to the data input of the latch element itself for a "hold" or "latch" operation.

As a result of the demand for increasingly faster designs, the synthesis time budget for module-to-module interfaces is ever diminishing. Fortunately, the shrinking geometries that enable the faster designs in the first place also mean that registers are more affordable in chip real-estate terms. Registers provide an approach to managing the diminishing module interconnect time budgets. As a general rule, module outputs should be registered where possible. This gives users of connecting modules driven by these outputs a greater time budget within which to synthesize. Optionally, registering of all or selected inputs to a module can also be considered. However, when combined with a strategy where all outputs are registered, additional input registering may be an over-the-top approach. Clearly, the pipeline delays associated with registering signals must be tolerable within the given design context to support registering of input or output signals. With certain time-response-critical circuits, it will be necessary to use a signal during the cycle in which it is generated, rather than waiting until the following cycle.

2.3.5 Use of Standard Internal Buses

For SoC applications or subsystem-on-a-chip applications, design reuse becomes easier if standard internal connection buses are used for interconnecting components of the design. Design teams developing modules intended for future reuse can then design interfaces to the standard bus around their particular module. In theory, this allows future designers simply to slot the reuse module into their new design, which is also based around the same standard bus, without the need to create wrappers around the interface.

2.3.6 CPU-Addressable Registers

Some modules may contain internal CPU-addressable registers. As explained above, the addresses should be defined using constants. The module registers will normally occupy a range of the CPU address space. Ideally, the address should be made up from a part that is a base address and a part that is an offset address. If the system address map changes in a later design, changing the base address in a central constants definition file may be all that is required to allow the same module to be reused.

The register access method should also be carefully designed. The number of chip clock cycles required to access registers may vary. For example, if a new design reusing a register-based module is interfaced to a slower microprocessor than that used with the original design, the number of access cycles may need to be extended. It is, therefore, important that a method for extending the number of access cycles be incorporated into the design. Furthermore, the method used should be consistent throughout the company.

2.3.7 Technology Specifics

Reusable modules should not instantiate technology-specific gates or macros. Clearly, this causes problems if the design is reused with a different target technology and the specific gate type is not supported or has considerably different characteristics in that technology. In cases where instantiation of technology-specific cells is necessary, the reasons for this requirement should be documented, and any associated limitations, restrictions or implications on future reuse should be noted.

One area where instantiation of technology-specific components can be unavoidable is in designs that use embedded memory cells. If future reuse of a module using a very specific memory configuration is a consideration, it is important to establish the probability that the vendor will continue to support the specific memory type being used in future technologies. However, better still, avoid using very vendor-specific memory cells. For the simpler memory types, there is a degree of commonality across the offerings from different technology vendors. Some advance consideration may allow the design to rework with a different vendor's memory cells with only minor tweaking of the design.

2.3.8 Verification, Testbenches and Debugging

In the ideal world, testbenches for reusable modules should be completely self-checking and should give 100% test coverage of the modules in question. In the real world, time-to-market pressures may not allow these goals to be met fully. Nevertheless, they are important goals and, therefore, should be given high priority. As is the case with design module code, the testbench code should be easily understood and adequately commented. Frequently, a module will be reused in a slightly different way than before or will require minor modifications. In such cases, the testbench may need corresponding changes, and it is important that the original testbench designer minimizes the effort required to make such changes.

Testbenches often execute tests under control of a test configuration file. These configuration files should be available for reuse and should be well commented to help when modifications are required.

Testbenches themselves are also ideally designed for reuse. For the same reasons that module reuse makes sense, designing testbenches for reuse also makes good sense.

Because one of the goals of design reuse is to enable a designer other than the original designer to reuse the design without necessarily understanding all the design details, it is a good idea to consider adding special debug features in more complex modules. By doing this, any design problems that do occur may be more easily discovered by a designer not fully aware of all the design details. Examples of debug features could include such techniques as the ability to bypass certain modules or submodules in a given mode, the ability to drive or preset critical inputs from registers or input pins, routing important internal signals or buses via a register-controlled multiplexer to output pins, logging of important internal debug signal to text I/O files, etc. Some debug features can increase simulation times, so all debug features should default to off.

2.3.9 VHDL versus Verilog

To approach the goal of a widely reusable module, it may be necessary to code the module in both VHDL and Verilog, because both these languages are in widespread use at the time of writing. If dual language support is seen as a reuse requirement, both the module code and the testbenches must be coded in both languages. If supporting both languages, it is important to avoid using keywords of one language while coding in the other language. A compiler will detect illegal use of a keyword in its own compilation language but will not be looking for use of keywords of other languages.

2.3.10 Live Documentation

When a module is first reused in another design, it is a good idea to encourage the reuser to create a reuse tips file. This should contain any additional information that, as a result of the reuse experience, the reuser considers useful but missing or incorrect in the original documentation. Once this file is created, it can be updated as appropriate on each reuse of the module. Any significant errors in the original documentation should, of course, be corrected.

2.3.11 Reviewing

Reviewing is an important part of the design flow. It is advisable that the code and documentation be reviewed by senior members of the design team, not only to verify its functionality, but also with a view to assessing ease of reuse of the modules.

A review of a module intended for reuse should arguably be more rigorous than the review of a use-once module. Any errors that slip through the review can create problems in many future chips, rather than just one.

2.4 SoC and Third-Party IP Integration

Up to this point, this chapter on reuse has considered reusability largely from the point of view of the developer of reusable modules. This section now looks at reusability from the point of view of the reuser and specifically from the point of view of a SoC developer reusing external third-party IP. Large SoC designs usually necessitate incorporation of third-party IP modules. In a sense, this is analogous to using standard off-the-shelf chips in traditional board design. SoC designs are typically comprised of a collection of relatively large independent functional building blocks interconnected via a system bus to realize a specific system configuration. Each of the functional building blocks should be considered by the project manager and design team by examining some of the criteria listed in the following section.

2.4.1 Developing In-House versus Sourcing Externally

When considering each of the modules required to build a SoC product, the project team must determine whether it is best to develop the module in-house or to source it externally. Asking some of the questions below will help the team to arrive at the correct decision.

- Does a module of this type or a sufficiently similar module already exist in previous in-house designs, and if it does exist, is it reuse-ready?
- Does the design team have the expertise to design the required module?
- Is there a strategic interest in developing this expertise in-house if it does not already exist, or is this a once-off requirement?
- Given resource constraints and time-to-market constraints, does it make more sense to buy in the required building block in the form of reusable third-party IP, thus freeing the design team's resources to focus on their own added-value specialties?
- What is the cost of the the third-party IP? Does it make economic sense to buy in?
- Is there a third-party IP module available on the market that matches the required functionality, or will it be a custom IP development?

It is worth considering some of these questions in further detail. The starting point in an analysis of this type should be an examination of the company's own IP library, if they have one. If they do not have a formal reuse methodology in place, it is advisable to proceed with caution. Although a similar IP block may have been developed in previous in-house projects, reusing this IP block may end up taking longer than designing one from scratch if the original requires some modifications but was not originally designed with reuse in mind.

If a sufficiently similar IP block is not available in-house or does exist but is not reusable, does the current design team have the expertise to develop a new one? A common example of where this is not the case is in the design of analog front-ends for what are otherwise digital chips. Analog design expertise tends to be the preserve of companies specializing in analog-only designs or in the analog design department of ASIC design houses. It is generally simply not an option to consider developing analog expertise from scratch. In cases such as this, there is no option but to buy in the required IP.

Assuming that a company has the required in-house expertise to develop a required IP block, the question of whether it is in the strategic interest of the company to develop this block must be asked. What if the field of expertise involved lies outside the core competence skills of the company? If such a block is to be used on a once- or twice-off basis, serious questions arise over whether it is worth investing time and effort to develop it in-house, even if there is a high degree of confidence that it can be successfully developed. Might it not be better to purchase this IP externally and concentrate on developing the added-value components that represent the company's core areas of expertise?

Let's assume that the expertise for developing a required IP block does lie within the company's core competence areas. This does not automatically mean that it makes sense to go ahead and develop it. Resource shortages combined with time-to-market pressures may dictate that the required IP is still sourced externally.

One of the most fundamental decision factors when considering third-party IP is the question of cost. The cost of buying in IP must not exceed a level that reduces the potential profit margin of the end product to an unacceptably low level. Consider also whether the IP can be reused in future projects. This lowers its effective unit cost if the purchase contract allows for unit cost reduction with increasing volumes.

2.4.2 Where to Source IP

If a decision is taken to source IP externally, where is the best place to find the IP? One of the first sources to consider is your regular ASIC vendor. Most ASIC vendors themselves have built up libraries of reusable IP to make their offerings more attractive to potential customers. This is often a combination of cores they have developed themselves combined with third-party IP that they have bought in or are licensed to use. One of the big attractions of using your ASIC vendor as the source of IP is that the IP will have been fabricated and tested in your target technology before (unless, of course, you are the first customer to use it, but even then, it is likely that the vendor will have tested it in advance). If the vendor is not the outright owner of the IP, a license payment per usage may apply. Alternatively, it may be built into the ASIC vendor's NRE charge. It is important to establish this early on to assess the impact on cost of the end product.

If your preferred vendor cannot source the IP functions that you are looking for, you obviously have to start looking elsewhere. There are numerous companies now scrambling to capitalize on the growing demand for IP. Many of these are new, small companies and may not be well known. One method of finding the IP you are looking for is through Web searching. There are already numerous Web sites up and running that specialize in matching IP providers with potential customers and vice versa. For example, at the time of writing, *http://www.siliconx.com*, *http://www.hellobrain.com*, and *http://www.design-reuse.com* all provide services of this type.

2.4.3 Reducing the Risk with Parallel Third-Party IP Development

If the IP block you require is very application-specific, there may be no standard off-the-shelf solution available, and you may have to engage a third-party design company to develop the required IP in parallel with your own design team's development of the rest of the chip. This is, of course, a risky option, but sometimes there are no alternatives. A number of basic steps can be taken to reduce the level of risk. Depending on the relationship with the third-party developer and on their geographic proximity, some of these steps may not be feasible. However, if they are feasible, it is advisable to consider taking these types of steps:

- Begin with a broad requirements specification but then try to involve the third party in developing the detailed specification. There is no point investing significant time in a detailed specification that simply may not be feasible.
- As with selecting an ASIC vendor, it is worth evaluating more than one IP provider with a view toward designing the IP in question. Preferably, the chosen provider will have worked with your company before and established a track record or at least have developed a reputation in the industry as a reliable IP developer. If this is not an option, closer monitoring of the IP provider's development program is mandatory.
- If possible, get the third-party IP developer to agree to the participation of senior members of your design team in its design reviews. If it is reluctant to do this and there are numerous possible providers capable of developing the required IP, it may be possible to make it a condition for getting the business from you.
- Ask the developer for bus-functional models as early as possible. It is probably better to use the developer's models rather than your own because, otherwise, you are simulating with what you think the developer is going to

provide, rather than with what the developer thinks it is going to provide. Any misunderstandings in the requirement specification will be reflected early in simulations if the third-party developer provides the models.

• Agree on frequent milestones with the developer and track these closely. If the development starts to slip early on, it may be necessary to have a contingency plan in place. In an extreme case, this may involve canceling the design contract at an early stage and awarding it to a competitor. Alternatively, it may be possible to source existing IP that, although not optimized for the required function, will at least meet the minimum requirements with some limited amount of redesign. This can act as a temporary solution while a more optimal solution for the original requirements is sorted out.

• Consider the size and experience of the developer's design team. The team should consist of some senior engineers, preferably with considerable expertise in the application area. It is important to establish that key team members will be working full time or nearly full time on the development program for its entire duration, rather than just being wheeled out to create an initial favorable impression to help win the business. The level of support needed to transfer the IP and a mechanism for implementing the transfer should be clearly understood and agreed on.

• Most important, the deliverables, costs, delivery dates and acceptance criteria should be documented and formally agreed on. These deliverables should include all work items, including specifications, design code, testbenches, simulation lists, design/implementation documentation, application notes, development tools and support activities. The third-party developer should work to the same high standards as the in-house design team to reduce project risks.

2.4.4 Issues with Third-Party IP

Using third-party IP is not always simple. You do not have the same level of knowledge or control as is the case with in-house developments. Here are just some of the basic issues to consider before embarking on this path.

2.4.4.1 Documentation

Does the IP come with adequate documentation? With in-house designs, you can often ask the designer if something is unclear or, at the very least, you have access to the source code and can try to figure it out yourself. With external IP, this is not the case. It is, therefore, vital that everything you need to know is contained in the documentation.

2.4.4.2 Interfaces

All interfaces should be examined closely. Are they compatible with your system? Are there any subtle tricky timing characteristics governing any of the interface signals? If the interfaces available do not match your requirements, how much effort and time is involved in developing bridging modules to adapt the interface into a form suitable for your system? Will such a bridging module introduce additional delay cycles that prevent your system from meeting its minimum performance requirements?

2.4.4.3 Testbench

The IP provider will often provide a testbench environment consisting of bus-functional models and monitors for testing its IP. However, components of this testbench environment may not map conveniently into your testbench environment. The testbench should be discussed in detail with the IP provider. It may be possible to get the provider to adapt testbench components to better fit into your environment.

2.4.4.4 Modifications

Modifications may be required if the available IP does not quite meet your requirements. Additionally, modifications may be required if previously unknown bugs are discovered. If it is a soft IP core, do you have access to the source code for making minor modifications, or do you have an agreement with the supplier that it will make any required modifications on your behalf? If it is a hard IP core, will the supplier tailor it to your specific requirements if it is not exactly what you require?

2.4.4.5 Testability

If an IP core is provided as a soft core, the core is then simply included as part of the greater system in terms of testability. Provided that it does not have many special test limitations, such as multiple clock domains or triggering on both edges of the clock, it is simply part of a larger chip from the point of view of the test strategy. For example, in a full-scan methodology, the scan insertion tool will not differentiate between one functional building block and another. It will not know—nor does it need to know—that one block was sourced externally and another developed in-house. However, if the core is in the form of hard IP, the test methodology is predetermined. Typically, hard IP will come with built-in full-scan ports or, alternatively, with a JTAG interface. Whatever the predetermined test methodology for the third-party IP, it will have to be integrated into the chip test methodology and will have a significant impact on the overall chip test methodology. It should, therefore, be considered in advance before purchasing the IP.

2.4.4.6 Layout

Does the IP block have any special layout requirements? If it is a large IP block, will you be allowed to run routing layers from surrounding blocks through it or will they have to run around the perimeter, thus resulting in potential knock-on layout problems for signals in other blocks? Again, this is an issue that should be discussed with the IP provider in advance of purchasing the IP. If your ASIC vendor is the provider of the IP block, it may have had experience of laying out the block in your target technology before and should, therefore, be aware of the main layout issues.

2.4.4.7 Layout with Deep Submicron Technology

SoC chips are often fabricated using the latest deep submicron technology, where routing delays dominate over gate delays. In this case, the IP should be provided as a hard core that includes the actual layout information or as a soft core, with synthesis scripts, layout constraints and guidelines. Layout information can be used when the design flow includes the latest synthesis tools that take account of layout information while optimizing the design.

2.4.5 Processor and DSP Cores

Processor and DSP cores are typically the most complex cores to integrate into a SoC from the point of view of system verification. Some form of co-verification of software and hardware is more or less essential when processors or DSPs are being used as third-party IP building blocks in the system. The time-to-market requirements of SoC products no longer allow hardware and software to be developed independently with an integrated verification phase only occurring after the independent development programs. Assuming that the IP core has been extensively tested by the provider, it is normal to use a C or C++ instruction set model to simulate its behavior in the system. C is normally used to model at a higher level of abstraction than VHDL or Verilog modeling and, therefore, allows a far greater number of tests to be executed. However, it is still advisable to run a limited sample number of tests with detailed low-level models.

Providers of processor and DSP cores will often be able to provide a hardware-based development system. Typically, this consists of an FPGA board with the embedded processor/DSP functionality. Ideally, it also comes with a verification suite that allows typical debugging features, such as instruction logging, single stepping, examination of internal registers, etc. Spare capacity on the FPGAs is then used for mapping in other system components that can be tested with real software and the target embedded processor/DSP. Software/DSP engineers should assess the software development tools provided by the third-party vendor and be involved in the decision to select an IP core.

2.5 System-Level Design Languages

Large SoC designs are acting as a major driving force in the generation of new tools and new system-level design languages. The new languages create "executable specifications" that allow quick analysis of different architectures and provide a design framework that supports cosimulation at many levels of abstraction. EDA vendors are vying with each other to provide a standardized language that will provide the design environment of the future, because the current VHDL/Verilog languages do not easily support higher-level modeling constructs. At the time of going to press, several languages are available from a number of consortiums, but none are currently standardized. Some EDA companies are extending VHDL and Verilog to provide system-level constructs, whereas others are using other languages, such as C^{++}, Java or specialized code. Whatever language is used, to get the most from this approach, traditional design flows will require significant modification, and design teams will require training to become proficient in the new methods of working.

2.6 Virtual Socket Interface Alliance

The Virtual Socket Interface Alliance (VSIA) is an industry group whose charter is to facilitate IP reuse. It promotes open interface standards that will allow "virtual components" to fit quickly into "virtual sockets" at both the functional level (e.g., interface protocols) and the physical level (e.g., clock, test and power structures). The VSIA was formed in 1996, and, at the time of writing, it has some 200 VSIA members, including individuals, companies and organizations. For more information on the VSIA, visit its Web site at *http://www.vsia.org*.

2.7 Summary

With ever-increasing gate counts and the advent of SoC designs, design reuse is becoming essential. Companies that avoid facing up to this challenge will face increasing and unrealistic development times on large designs. Increasing the size of the design team is not only a costly approach to reducing development times, it simply may not produce the desired end result. There comes a point where modules cannot sensibly or efficiently be subdivided into smaller blocks for distribution to additional engineers. Adding more engineers at this stage achieves little, if any, benefit and increases the overhead on the project manager.

The degree of success in adopting a reuse philosophy can be considerably enhanced by designing with reuse in mind from the outset. A number of practical tips for improving the reusability of a design have been given in this chapter. Arguably, the most important two things to facilitate future design reuse are the provision of comprehensive design documentation and a comprehensive testbench.

From the point of view of the SoC developer, there are numerous considerations to take into account before sourcing IP for reuse. A number of these have been examined in the latter part of this chapter.

For many individuals and organizations, a cultural shift is required to embrace fully the philosophy of design reuse. It is the role of management and project leaders to introduce a design reuse culture successfully by explaining the necessity for it and by providing appropriate motivation and rewards for design teams that successfully take design reuse on board.

## CHAPTER 3

A Quality Design Approach

3.1 Introduction

The primary objective of the project team is to deliver a fully functioning ASIC within defined timescales. Without a quality approach, there is a considerable risk that the project will result in failure. A high-quality approach will pay dividends for current and future projects, because many future projects often end up using modules from current ASICs. The project structure, which defines such things as revision control of documents, organization of files in the directory structure, fault-tracking, etc., is discussed in Chapter 7, "Quality Framework." This chapter is focused on a quality design approach.

A quality design approach will reduce problems as the project progresses, especially around the chip integration time. When quality is designed in, modules bolt together seamlessly, without the need for reworking. Chip integration is a key time because any slip against the plan at this advanced stage of the development normally means a slip to the product end date.

The focus on quality must begin at the start of the project. The need for quality should be impressed on the team. Poor quality can lead to major slips to the plan or silicon that does not function correctly, which, in turn, leads to respins.

The top-down approach yields a well-constructed project. This means defining requirements before designing the architecture and designing the architecture before working on the lower-level modules. However, some of the lower-level critical modules may need to be prototyped to check assumptions when designing the architecture.

All important project documents that are generated should be reviewed, maintained under a formal revision control system and kept up to date. Marketing and management should sign the project or ASIC requirements document. Any changes to the requirements document after initial specification should be signed off by all relevant parties.

The following sections describe different aspects of a quality design approach: design documentation, reviewing, a quality approach to the module design phase and quality top-level simulations. The final section provides some suggestions for review checklists.

3.2 Project Design Documentation

3.2.1 Specifications

There is always a temptation to start the design before the specification has been agreed on. In the ideal project sequence, some form of project specification is available, from which the design team writes an architectural specification. For ASICs with some degree of programmability, the design team should also write a programmer's reference specification early in the project cycle. This should contain register addresses and content field descriptions (which explain the meaning of the register bits). It should also provide examples of programming sequences to set the chip in various different operational modes and, if necessary, some text descriptions of the associated hardware to explain the context of the registers.

It is extremely important that the architecture document and programmer's reference specifications are updated regularly with any changes. These documents are vital so that everyone can understand the project properly. Maintaining these documents in a clear, concise form and keeping them up to date will help to ensure that the modules designed by the various individual designers and design teams will provide the desired end functionality, when integrated.

3.2.2 Hierarchy Diagram

The top-down design approach begins by breaking the ASIC down into a number of top-level modules. For example, a communications chip might logically partition into a transmit block, a receive block, a physical interface block, a processor interface block, a memory management block, etc. Having agreed on the partitioning of the top level, a hierarchy diagram should be created. This defines the interconnections between the modules. The hierarchy diagram should be accompanied by a "data-dictionary" document that defines the signal names, characteristics (e.g., a single-cycle pulse, active-high, with period P), bus widths and functional description. This technique should be applied recursively down through the design, breaking up each new top-level module into a set of submodules until the submodules do not logically divide down any further and are of a manageable design size.

During any partitioning exercise, modules should be partitioned by functionality, rather than size. Conceptually, it is easier for designers to design modules with a unified function, rather than groups of miscellaneous functions. The same is true when it comes to reviewing designs. A further reason for partitioning by function is that the likelihood that a module can be reused at a future date is much higher if the module performs a single function or a set of closely related functions.

The ASIC vendor may have guidelines for the number of top-level modules. Some vendors allow simple floor planning using the top-level modules, and some of these tools are limited in the number of modules they can practically handle. It is difficult to generalize about the correct number of submodules to have at any given level of the hierarchy. However, as an arbitrary guideline, it may be useful to aim for somewhere between three and seven modules. As this upper level is exceeded, it becomes progressively more difficult to understand the functionality by visual inspection of the hierarchy diagram. Additionally, the number of interconnects required may clutter the diagram to the level where it becomes confusing.

The hierarchy diagrams are very useful for checking interfaces between the modules. They are a valuable aid in the top-down design approach. In combination with the data-dictionaries, they make the chip easier to understand for other engineers (including manufacturing and test) and are a valuable aid when it comes to subsequent design reuse.

3.2.3 Design Route Document

Many design companies support the use of a range of ASIC design tools. The design route document should define which tools—and which revision of each tool—should be used during the project. The document should identify any templates that should be used for particular tools (for example, the location of a common synthesis script). The document provides a good location for tips and guidelines for the particular tools. Any known tool or design process issues should be highlighted in this document.

The design route document should be reviewed periodically to ensure that it reflects the latest decisions on tools and design environment, and to ensure that everyone is working in an identical design environment. In addition to covering tools and design environment, the document can include guidelines on design documentation and decisions on test strategy.

3.2.4 Module Documentation

Design documentation for ASIC modules is important so that other engineers can review the design and so that software and test engineers can develop code in parallel with the ASIC design. Documentation is a necessary part of a professional approach to the design. Each module should have a module specification that includes a definition of the function of the module, register descriptions, a module block diagram (hierarchy diagram) and interface descriptions (data-dictionary). During the design phase of the module, the designer should generate further documentation, including submodule hierarchy diagrams, block diagrams, flow charts (where applicable), timing diagrams and design notes.

Any module that is considered to be a higher-risk module—perhaps because it is more complex or is being designed by a less experienced engineer—should have more in-depth documentation. As a general guideline, design notes need not contain huge amounts of text describing the precise function of each part of the module. It is preferable that there is just enough to capture the essence of the functions, along with descriptions of any special design issues. Some of the diagrams may not be in electronic form. Electronic form is preferable from the point of view of archiving, but it is not essential, as long as the diagrams can be understood during reviews.

When a module has been designed and fully tested, it is useful to update the documentation with an implementation document that describes the final design.

A further important part of module documentation is the module simulation specification. This should be written at the start of the simulation phase. However, it can be useful to start thinking about simulation before completing the design, because it may prompt the designer to think twice about creating module architectures that are unnecessarily difficult or time-consuming to simulate.

3.2.5 The Test Approach Document

The test approach document describes how the ASIC will be tested. It explains any special test structures and covers all test issues. A chip is worthless if it cannot be tested sufficiently. The test approach document covers two types of test. The first type enables the ASIC vendor to verify that the chip has been fabricated correctly (no short-circuits, stuck-at-faults, etc.); the second type enables manufacturing to prove that the chip is working when added to a PCB. The document should define the minimum fault coverage percentage (this coverage figure is typically generated by the test tools) and is typically over 90–95%.

The test approach document should be reviewed and accepted by the software, test and manufacturing teams.

3.3 Reviews

Reviewing provides the foundations of a high-quality project. The project leader should stress that the objective of reviewing is not to criticize but to identify problems at an early stage of the project. The way in which reviews are conducted affects the way in which designers respond to the reviews. If reviews are run in a positive, constructive manner, designers will gladly participate in them.

It is important at the start of the project that the project leader defines what will be reviewed, when it will be reviewed, and the objectives or reasons for the review. Ideally, for each review, there will be a checklist of items to review (some suggestions for checklists are given at the end of the chapter). This allows designers to prepare better for the review.

The project leader should also identify who should be involved in the reviews. Ideally, the system architect and the project leader should always be involved. The system architect must ensure that the module works within the system, and the project leader ensures that actions are recorded and tracked, and that the review has a motivating effect on the designer. Other team members can be involved in the review for part or all of the review. For example, when reviewing the interface to another module, that module owner can be brought into the meeting. Care must be taken in deciding who should do the reviewing and in conducting the review itself, because designers (especially inexperienced ones) can easily assume that they are being criticized. Normally, a small number of senior designers makes a good review team. Sometimes, it is just the system architect and the project leader.

If conducted properly, reviewing can have a positive motivating effect on a designer. However, if handled badly, it can have the opposite effect. The project leader should highlight good work and give some positive overview comments at the start of the review. It is useful for the code reviewers to analyze the code prior to the meeting, so that reviews run quickly.

If a module is to be reused, extra reviewing is necessary at each stage. The reviewing should establish that the design is in conformance with any company specific guidelines for reuse. Reuse is covered in detail in Chapter 2, "Design Reuse and System-on-a-Chip Designs."

3.3.1 Review of the Architecture Specification/Register Specifications

This should involve the entire team, including software, test and manufacturing engineers. Ideally, it should also include engineers from other groups and projects. The reviewing process is a good mechanism for enabling engineers to understand the system in its entirety and to understand how the individual modules that they are working on relate to other modules. This will help them to put their designs in context and to understand better the real purpose of their modules in the overall design context. Any such improved understanding enhances the chances of getting the design right the first time and usually has a positive motivating effect on team members.

3.4 Module Design and Reviewing

A quality approach to the design of a module can be broken down into a number of steps: specification, design phase, coding, simulation/synthesis and documentation. These steps are illustrated in Figure 3-1. A top-down methodology is consistent with the concept of a quality approach. It offers the prospect of a high-quality result with the lowest risk of slips to the plan.

3.4.1 Module Specification

Before the design of a module can begin, a clear specification is required. Part of this should be included in the architecture document. The module designer should work with the system architect to generate a module specification.

This specification should be reviewed at the start of the design phase with an aim that the module will work when integrated into the system. It should be reviewed with respect to each of the following headings:

Functionality: The functionality should be defined in a clear form. This can take the form of text, "pseudocode," diagrams, C programs, etc. The specification should include the time taken to complete various operations (e.g., it will take n clock cycles to generate a new data word) and the throughput rate. It should state any assumptions that are made. It should identify any conditions where the functionality will fail and the effect of the failure.

Module I/O: The I/O signals should be defined using sensible names that give some indication of the signal function without being excessively long. These names will be the names used in the VHDL or Verilog code and, therefore, must not contain any invalid characters. The project or company coding rules should give a guideline for

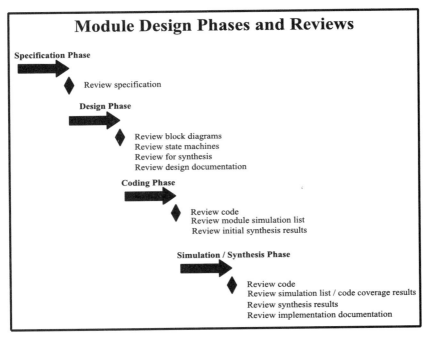

Figure 3-1   Module Design Phase

forming names (e.g., first letter capitalization as in *DataValid* or use of underscore characters, as in *data_valid*), and this naming convention should be adhered to consistently in the specification.

Timing Information: The specification should also provide timing diagrams in cases where the timing is complex. Ideally, each I/O signal should conform to a project standard timing requirement. A project standard timing requirement could be that all outputs are valid *x* nanoseconds after the system clock, and all inputs are set up and stable *y* nanoseconds before the system clock. The sum of the *x* + *y* timings should be less than the system clock period to allow for clock skew and extra delays after layout. The specification should identify those signals that do not conform to the standard timing requirement and clearly define their unique timing characteristics. If a module owner is using different timing requirements than the project standard for IO signals, the non-standard timings must be agreed on by the other module owners affected by the non-standard timings.

Internal Registers: All registers should be clearly defined. The definitions should include the register address, the function of each bit or field in the register, the power-on reset or default values of the bits and the assertion level of each of the bits. Wherever possible, write-only registers should be avoided, and clear-on-read registers should be used with care. Write-only registers are something of a limitation when it comes to debugging—a successful write operation cannot be confirmed by simply reading back the written value. Clear-on-read registers or toggle-on-read registers often cause confusion for programmers. The software, test and manufacturing engineers should be involved in the register reviewing. The register definitions should form part of the programmer's reference guide. Each module owner should use a standard format for representing registers so that the documentation style is consistent.

Test and Debugging Modes: The specification should describe any special test and debug modes. These could include modes in which signals can be multiplexed out to the top level when the silicon is being tested, descriptions of parameters that can be modified or configurations that can be selected if the ASIC is not performing as expected (e.g., the percentage full setting for a FIFO at which new data is requested).

Some modules will benefit from test registers that configure the module to a certain state that allows easier testing. For instance, in a simulation that takes a long time to fill a FIFO to a critical almost-full condition, it can be useful to have a test register that artificially makes the FIFO appear almost full immediately.

3.4.2 Module Design Phase

The design phase is the most creative part of the module design. The objective is to create a design that will fully implement the specification. The output of the design phase is normally in the form of documentation, such as hierarchy and block diagrams, state machine diagrams, timing diagrams and implementation notes. Well-thought-out documentation will minimize the time needed for the coding stage and make the coding task easier. Sometimes, it is appropriate to generate a number of design options for a particular function. These options can then be reviewed and the best approach selected.

The documentation should be reviewed so that system problems and implementation errors are highlighted before coding and testing starts. The review should also consider the impact of the design architecture with regard to synthesis. For example, some architectures may be less likely to work at the correct speed or power, or will result in excessive gate counts. Any potential critical timing paths should be identified and, in some cases, some test code generated and synthesized. For example, the design may use a number of complex adders in series. A test case will show the maximum speed of the logic. Such a design approach is an example of using a combined top-down/bottom-up approach.

At the initial design stage, not all documentation needs to be in electronic format.

The project leader should ensure that designers do not spend excessive time generating diagrams (unless they are useful for the coding stage) because the design is more susceptible to change at this early stage. The design may change as new requirements are added, as a result of the design review, or after initial simulations. Such changes may make the some of the design notes redundant.

The design review should be held before coding starts, and should check the following:

Hierarchy and Block Diagrams: The block and signal names should be sensible and, where possible, match the names in the specification. Typically, each block will be coded as a separate VHDL or Verilog file. The file name should match the block name. The review should ensure that the module names are unique. In some modules, a number of blocks can be coded into one file. In this case, the block diagram should identify which blocks will be in which files.

Synthesis Issues: The synthesis issues of the design should be reviewed. Block interconnect signals should have their timing budgets defined. The object of this part of the review is to force the engineer to consider the synthesis approach from the start of the design. Some synthesis tools can create different architectures for arithmetic operations, such as addition and multiplication. The analysis of which architecture is chosen should be carried out and reviewed during the design phase.

Timing Diagrams: The timing diagrams should be analyzed. Particular attention should be given to unusual timing windows, e.g., an extra timing cycle required when the data takes on a particular value or range of values.

State Machines: Some modules will contain state machines. These should be drawn up in diagrammatic form before coding. The review should check for correct functionality and for deadlocks in the state machine (a state in which it can hang, because some event has happened at a different time than expected). Ideally, the state names used in the state diagrams should be the names used in the VHDL or Verilog code. In VHDL, the state can be defined as an enumerated type. Care must be taken since in RTL (register transfer level) code, enumerated types will always assume a default value of *LEFT* until they are explicitly defined. This will not happen in the gate level. Signals that are used to trigger an edge-sensitive device should have registered outputs. Signals that cross clock boundaries should be highlighted. The review should analyze the time required for the state machine to perform its various operations and should check that this is within the time allowed in the specification.

Implementation Notes: As part of the design phase, the designer may generate some implementation notes, which will be used during the coding phase. These could include analysis of latency and throughput, required bit precision for arithmetic operations, etc. These should be analyzed so that the designer can benefit from the experience and analysis of other team members. Any design assumptions should be defined in the implementation notes and examined during the review.

3.4.3 Module Coding Phase

During this phase, the designer will convert the design documentation into VHDL or Verilog code. The coding should be based on standard coding guidelines. At the end of this phase, a module test list document should also be written.

A review should be scheduled at the end of the coding phase or after some very simple testing. A review at this stage is very useful because changes can be made to the code without significantly affecting module timescales (e.g., if a small section is coded inefficiently, it can be recoded quickly without wasting too much time). If the code review is not held until after testing, parts of the time-consuming testing process have to be repeated. Common errors, such as incomplete sensitivity lists or incomplete case statements, are often easily spotted during these reviews.

The code should be reviewed for ease of understanding. It is a good idea to make the assertion level of all signals active high for consistency because this is the convention adopted in most designs. The signals should have sensible names that indicate their function. If the ASIC is being designed using a top-down approach, the signal names at the higher levels of hierarchy are already predefined, and unless there is good reason for changing them at a lower level, the lower levels should inherit the same names. The code should be adequately commented. Each module and process should have a brief description of its function. The declarations for interface and internal signals should be grouped according to function, with comments where appropriate. The signal names should match the design documentation.

The function of the code should be analyzed. The code should be implementing what was defined in the design documentation. Particular attention should be given to the interfaces with other modules and especially to chip external interfaces. Arithmetic operations should be checked for bit precision and the handling of overflows. Consideration should be given to the inclusion of test configuration modes. The ability to disable or bypass some functions can be useful.

The code should be reviewed for its handling of different clock domains. Where necessary, signals should be double-registered to prevent metastability problems.

Test and simulation issues must be analyzed in the module. This can require direct access and control from external pins or it can require special test modes. The test modes should be designed for both chip testing and product testing. The module should conform to project test guidelines, and the review should ensure that the guidelines are followed.

Part of the review should focus on synthesis issues. The most important of these are covered in a checklist at the end of this chapter. An initial trial synthesis should be carried out and the resulting gate-counts and timings analyzed. The critical paths should be identified. All embedded synthesis directives should be checked (e.g., *sync_set_reset* attributes, synthesis off/on, etc.).

Access to registers should be reviewed. Wherever possible, they should be accessible in both read and write mode. This allows for a very simple form of testing, i.e., can written data be read back correctly from the same address? The addresses should be coded using constants with meaningful names, e.g., *ctrl_reg_addr* (ideally defined in a central constant definitions file).

Structures such as state machines should be reviewed for ease of understanding. The state coding style (e.g., gray, one-hot, random, etc.) should be discussed. Some synthesis tools will automatically create different state coding styles if appropriate directives are added.

Code analysis tools are available that highlight characteristics of the code. If a code analysis tool is available, the review should examine the output of the code analyzer.

The final part of a coding review should examine the list of simulations for testing the module. It is important that the list gives reasonably detailed descriptions of the tests. The reason for reviewing this list is to encourage a structured approach to the testing and to establish that the test list is sufficiently comprehensive. The list is also useful for the project leader because tracking the list of tests gives an indication of the status of the testing.

3.4.4 Module Simulation/Synthesis Phase

During this phase, the module is fully tested and synthesized to the required speed. The module simulation test list document should be updated with the results. Once the design is tested to the point where it is unlikely to change significantly, a design implementation document should be generated.

At the end of this phase, the corresponding review should concentrate on the functionality of the code, the quality of the testing and the synthesis results.

The synthesis results and constraints should be analyzed. The input timing and drive strengths, output loads and output timing should all be checked and any multi-cycle paths examined carefully.

The list of tests and test results should be reviewed and any errors or problems discussed. Test coverage tools are available that graphically show the test coverage of the module. This gives an indication of how much of the code has been exercised by the tests that have been applied.

3.5 Quality System Simulations

The success of an ASIC is heavily dependent on the quality of the system simulations. Ideally, top-level simulation results should be compared against a reference model, which is generated by a C program, Matlab or other such programs and tools. If the company has a research department, the reference model can be the method of transferring knowledge from the researchers to the design engineers.

The best approach to quality system simulations is to involve software engineers in the simulations. This ensures that the chip is tested in the way it will be used. This approach is described separately in Chapter 5, "ASIC Simulation and Testbenches," which also describes the characteristics of a high-quality testbench.

3.6 Review Checklists

Reviewing is more likely to be effective if a checklist is available before and during the review meetings. The checklists are useful for the designers because they are aware in advance of what the review will address. The following subsections give some suggestions for different review checklists. How applicable the lists are will depend on the design type, but many of the checklist items are common to a wide range of designs. A number of items appear in more than one of the checklists. However, this simply reflects the fact that certain fundamental issues need to be checked in more than one of the review types.

3.6.1 Specification Review Checklist

- Review for correct functionality—does this specification accurately capture the requirements?
- Review throughput and latency.
- Review all assumptions made about the design.
- Review the limitations of the module—what it can do and what it cannot do.
- Review possible failure modes—what happens if certain parts do not work as intended, e.g., does the design lock, does it temporarily lose data, etc.?
- Review I/O signals. Check for any missing inputs or outputs. I/O names should be descriptive but not too long and should observe the relevant

VHDL or Verilog rules. They should also match the names captured in the hierarchy diagrams.

- Review interface-timing diagrams.
- Review register definitions. Check assertion levels of register bits, reset states and register addresses.
- Review for adequate test and debug support features.

3.6.2 Design Documentation Review Checklist

The main goal of this review is to encourage the designer to design the module before implementing or coding it.

- Review the hierarchy and block diagrams. The signal names should match the specification signal names. The block names should be the same as the corresponding names of the VHDL modules and files—ensure that they are unique.
- Review any timing diagrams. Timing diagrams should be drawn for functions involving complex timing cycles before coding starts.
- State machines should be reviewed for functionality. Check that the state machines cannot get permanently stuck in any particular state. Check the reset conditions of the state bits and all registered outputs.
- Review the implementation notes. Analysis should be carried out on data throughput, latency, FIFO depths, memory access bandwidth, etc. DSP functions should be defined (e.g., filter functions, etc.). Bit precisions, jitter, error analysis and allowable error rates, etc., should be examined.
- Review the use of clocks within the module and any intended timing relationships between different clocks.
- Review the handling of asynchronous signals entering synchronous clock domains.
- Review the test approach. Do any modules need to include specific test modes?
- Review for synthesis. Consider the critical timing paths, estimated gate-count and power consumption issues. Has a test case been generated? What were the synthesis results?

3.6.3 Coding Phase Review Checklist

- Review the functionality of the code against the module specification. Pay particular attention to sections that will be difficult to validate during simulation and as a consequence will rely on the code review itself to a greater extent than other sections.

- Review all I/O interfaces, both to other on-chip modules and, particularly, to external interfaces.
- Review the code for ease of understanding. The code should be adequately commented. Signal assertion levels should be consistent, preferably active-high because this is the general convention. If the code does make use of active-low signals, this should be signified in the signal names, e.g., through the use of an _n or _l suffix. Signal names should give some indication of the function of the signal. Ensure that the same signal name is not used at higher levels in the design.
- Review any arithmetic operations. Check bit precision and overflow handling.
- Review the handling of multiple clock domains.
- Review test and debug issues. What signals will be available for debugging when the ASIC is being tested on the bench? Are there any simulation test modes that will help the validation of this or other modules (e.g., a test mode that allows a FIFO to be initially part prefilled so that it is easier to check the overflow condition)?
- Review the synthesis issues. Check for unnecessary *if...else if* priority encode constructs, unwanted inferred latches, behavioral code that will simulate differently from the gate-level code (signals not reset correctly, enumerated types powering up differently, signals missing in sensitivity lists, etc.). Check that any necessary synthesis directives have been included (*sync_set_resets*, synthesis off/on, etc.)
- Review the results of any trial synthesis. The gate-count and worst-case timings should be analyzed. Check that the results are sufficiently close to the requirements that minor design modifications or better synthesis scripts will be enough to allow the synthesis requirements to be met fully.
- Review the usage and coding of all programmable registers. Ideally, all registers should be readable and writable. Addresses should be defined as constants in a central file.
- Review the simulation test list for comprehensive simulation coverage.

3.7 Summary

This chapter has suggested methods of ensuring a quality design approach. The importance of design documentation has been stressed, and suggestions on the essential contents of a documentation package have been presented. A good documentation package is all the more important if it is likely that the design will be reused in the future.

There has been a strong emphasis on the review procedure, with suggested reviews occurring after the specification stage, the design phase, the coding phase and the simulation and synthesis phase. The purpose of each of these reviews has been explained, and a suggested checklist has been given to help identify the common types of design mistakes that are encountered in the above phases.

## CHAPTER 4

Tips and Guidelines

4.1 Introduction

This chapter covers a number of guidelines that should serve as a help to experienced and less experienced designers alike. The guidelines covered are by no means exhaustive. They do, however, cover many of the common issues that arise over a wide range of design types. Although design content may vary enormously from chip to chip, many of the problems encountered remain the same, and many of the methods of avoiding these problems are applicable across multiple design types. Companies involved in ASIC design should have their own published design guidelines that are reviewed and updated periodically to ensure that they remain up to date and reflect what are regarded as industry best practices. Some of the advice in this chapter inevitably overlaps topics covered in other chapters, such as the chapters on design reuse, synthesis and a quality design approach. This is done so that it is easier for an inexperienced designer starting out looking for some initial guidelines to read through one focused chapter, rather than having to search the entire text of the book.

The guidelines covered here fall into several categories. The initial general section covers style and design practice issues. Subsequent sections cover synthesis and test topics, and the last section provides an introduction to some of the issues involved in dealing with multiple clock domains.

4.2 General Coding Guidelines

As stated above, it is advisable that companies maintain a set of design guidelines.

Coding guidelines are especially important. If designers more or less stick to the coding guidelines, poor coding styles should be avoided, and designers should be able to understand, review and reuse other designers' code more easily.

This section lists a miscellaneous set of coding guidelines. It is difficult to categorize these in any particular manner or sequence because they cover a wide range of issues, and so the sequence in which they are presented is not significant. Code coverage tools and code-checking tools will often identify cases where good coding guidelines have not been adhered to. Note that the chapter on design reuse also covers the topic of coding guidelines but from a more specific perspective.

4.2.1 Simplicity

Keep the code and hardware structures as simple as possible to achieve the required function. If the design can be implemented simply, do it simply. Avoid unnecessarily complex code. The less experienced engineer may unwittingly think that complex code will impress peers and the project leader. Inexperienced designers may opt for wildly elaborate hardware solutions when a much simpler, safer solution is possible. Complex code is more difficult and more time-consuming to review, and, therefore, more prone to error. It is also less suitable for design reuse. Unnecessarily complex hardware architectures are also more prone to the introduction of design errors.

When complex code is unavoidable, ensure that it is adequately commented.

4.2.2 User-Defined Types and Complex Data Types

Avoid an excessive number of user-defined types. If someone other than the original designer is either testing the code or reusing it, they will have to examine the type definitions repeatedly to understand the code. Try also to limit the number of special purpose files/packages in which these types are defined because, once again, someone other than the original designer who is trying to understand the code will spend unnecessary time searching for unnecessary type definitions. In cases where user-defined types are particularly useful, such as states of a state machine, the types can be defined in the source file, rather than a separate package.

Avoid excessive use of complex data types, such as VHDL records, especially at block interfaces. On the one hand, these make the interface more compact from the point of view of coding and schematics. On the other hand, they reduce the amount of information that can be instantly acquired by simply examining the interface signal names. For example, contrast an interface that only consists of a record named *memory_signals_record* with an interface that consists of the individual set of signals that make up that record: *address(31:0), dataIn(16:0), dataOut(16:0), writeEn-*

able(1:0), *ramClk*, *outputEnable(1:0)*. The latter individual set of signals gives more instant information. Neither is there any need to go on a "grep" hunt of where *memory_signals_record* is defined.

4.2.3 Naming Conventions

Adopt a standard strategy for naming variables and use names that intuitively indicate the function of the signal. Two commonly used naming styles involve dividing the names with underscore characters or leading uppercase characters. For example, *en_burst_write_addr* or *EnBurstWriteAddr* are both more intuitive than *enburstwriteaddr*. The common strategy could also define an abbreviated naming style for commonly used key words that appear in variables. For example, *address* could be represented by *addr*, *write* could be represented by *wr*, *read* by *rd*, *enable* by *en*, etc. Adopting an abbreviated naming style avoids having painfully long variable names, and making it common should make the meaning instantly recognizable. A further specific instance where a variable naming strategy is useful is for the outputs of synchronizing registers. It is often useful to be able to identify these quickly in a netlist. For gate-level simulation purposes, we may wish to identify them to disable *X*-propagation. For synthesis purposes, we may wish to identify them to set false paths to them. A possible naming convention to achieve this is to prefix or suffix the normal register output name with the string *sync_* or *_sync*. Most synthesis tools will use the output name in the register instance name, thus making it easy to identify in the gate-level netlist.

Do not change the name of a given signal as it descends or ascends through the hierarchy. A common example of this is the use of several naming variations on the same clock signal throughout the design. *Designer A* names the clock *sysClock*, *designer B* names the same clock in a different block *systemClock* and *designer C* names it both *sysClk* and *mainClk*. This results in lots of confusion and lost time trying to reconcile names when it comes to system-level simulation and debugging. To avoid this, the design team needs to agree on names of common interface signals before module design begins—a simple but unfortunately often neglected step. If there is a good reason to rename a clock signal in certain blocks as it descends through the hierarchy, it should be renamed in a port mapping, rather than by using an explicit signal assignment in an executable statement in the module code. Signal assignments in executable statements result in delta delay cycles in the simulator. This causes a shift in clock timings and can result in failing simulations because clocks that the designer thinks are transitioning at precisely the same instant are in fact transitioning at slightly different times in the simulation environment.

File names should be the same as module or entity names, and, as a general rule, instance names should also be similar to module names and consistent in style. It is much easier to find relevant modules when this is done. For example, if we see a module named *statistics* instantiated in the *data_logging* block, it is much easier to guess that the source code for this module resides in a file named *statistics.v*, rather than in a file named, for example, *module1.v*.

If instance names are consistent in style, it is much easier to specify paths when searching for a specific instantiation in a file or when interactively using the synthesis tool to specify hierarchically a path or module. For example, if we are familiar with our design and know that the *bit_transmit* block lies under the *data_framing_block* which, in turn, lies under the *output* block, it is easier to remember a consistently named path such as

```
inst_output/inst_data_framing/inst_bit_transmit/
bit_count_reg/D
```

than it is to remember some mixed style, such as

```
output0/data_framing/bit_transmit_instance/
inst_bit_count_reg/D.
```

4.2.4 Constants

Use constants, rather than coding fixed numbers in the code. This makes the code easier to modify and understand. Use all uppercase characters for constants. This makes them instantly recognizable as constants (provided that variables are not named using the same convention!).

4.2.5 Use of Comments

Comments in the code can greatly help anyone examining the code (including the original designer at a later date!) to understand what the design is trying to do. Commenting should begin with a file header at the top of each file, showing the history of changes to the code, the corresponding version numbers, and the author associated with each change. The header can also include a brief summary of the overall function of the code, known limitations, lists of related modules, etc.

Within the code itself, further comments should be added at different levels. Unless the function of a process, function, procedure, etc., is self-evident, it is usually worthwhile adding a one- or two-line comment indicating the function of the process/ function/procedure. At a lower level, it is sometimes worth commenting individual lines of code. For example, in the VHDL line below, it is immediately obvious, with the aid of the comment, that the packet length field of some data structure is being cleared. Without the comment, the line does not have that instantly recognizable meaning.

```
RamDataOut <= ramDataIn(16 downto 8) & "00000000";
                              --Clear the pkt.length field
```

The code should not, however, be overcommented. Detailed descriptions belong in the accompanying design documentation, rather than in the source files.

4.2.6 Indentation

Code is more readable and more readily understood if an indentation scheme is used. It is especially useful in IF-ELSE-ELSIF clauses because it makes identifying which IFs, ELSIFs, ELSEs and END-IFs belong together. Contrast Example A below, which has no indentation, with Example B, which uses a sensible indentation scheme. With this particular scheme, it is easier to associate related IFs, ELSIFs, ELSEs and END-IFs in Example B, because those that are related have the same level of indentation.

4.2.6.1 Example A

```
if (sel = "00")then
regEnable <= "0001";
if (dataValid = '0')then
ramAddress <= txAddress;
next_state <= LOAD_TX;
else
ramAddress <= rxAddress;
next_state <= LOAD_TX;
end if;
elsif (sel = "01")then
regEnable <= "1001";
ramAddress <= (others => '0');
next_state <= LOAD_TX;
else
null;  --hold current values
end if;
```

4.2.6.2 Example B

```
if (sel = "00")then
    regEnable <= "0001";
    if (dataValid = '0')then
        ramAddress <= txAddress;
        next_state <= LOAD_TX;
    else
```

```
                    ramAddress <= rxAddress;
                    next_state <= LOAD_RX;
             end if;
      elsif (sel = "01")then
              regEnable <= "1001";
              ramAddress <= (others => '0');
              next_state <= WAIT_ADDRESS;
      else
              null;  --hold current values
      end if;
```

4.2.7 Ordering I/O Signals and Variable Declarations

When listing inputs and outputs to a block in a block header, it helps to use a consistent scheme. There are many possibilities, and here are some examples:

- Mix inputs and outputs together, but keep in alphabetical order.
- Split inputs and outputs into two separate groups, but apply alphabetical ordering in each of the groups.
- Arrange the signals in groups by function.

Similarly, lists of internal signal declarations should be ordered in some consistent way. The most common approach is simply to list them alphabetically. In large modules with large numbers of signals, it takes less time to find specific signals when ordered in a standard way.

4.2.8 Consistent Logic Assertion Levels

Try to be consistent in the use of logic assertion levels internally in the ASIC, i.e., use either positive or negative logic throughout, but avoid mixing the two. Positive logic means that control signals are considered asserted when "high." By sticking with a consistent assertion level (preferably positive logic, because this is the most popular convention), there is less possibility of misinterpreting assertion levels at module interfaces. In a positive-logic ASIC that has to drive active low external connections, invert the internal signal at the top level of the chip or in an inverting output buffer.

Sometimes, it is not possible to be completely consistent in the use of assertion levels. For example, in a positive-logic ASIC, the design may be using a predesigned macrocell that requires a mix of active-low and active-high inputs. In such a case, add a suffix such as _n to the active-low signal names to highlight the fact that they are active-low.

4.2.9 Use of Hierarchy and Design Partitioning

Design partitioning and the use of logic hierarchy arise as a result of necessity. For example, certain tools cannot handle excessively large, flat designs. However, even if it were not for tool limitations, we would still partition designs for reasons of design management and ease of understanding.

Synthesis tools cannot handle arbitrarily large blocks. Million-gate ASICs must be partitioned and systematically subdivided into blocks of manageable size. Tools such as Design Compiler have no hard-and-fast internal limit. However, their compile time increases exponentially with the gate count.

Floor-planning tools usually start with an initial set of block placements, where the initial blocks are the logical design partitions at some level of the hierarchy. Without block definitions and partitioning, there is nothing to base an initial placement on, other than the individual cells in a flattened or nonhierarchical design. Design partitioning is, therefore, a natural step in floor planning.

Hierarchy and design partitioning also make a design intuitively easier to understand. If a section of random logic contains more than 10 logic gates, it can become difficult to understand a flat schematic representation. A schematic with several hundred basic logic gates is almost impossible for a human to understand unless it is arranged in very simple parallel structures. However, if such a design is presented as small hierarchical logical groupings of functions such as accumulators, registers, adders, etc., its function then becomes much more intuitive. This argument scales up to the top levels of a chip. If we see the top level of a communications chip broken into hierarchical blocks, such as a receiver block, a transmitter block, a statistics block, a switching center, an embedded processor block, etc., we can instantly tell quite a bit about the design and know where to start looking for specific types of functions. Clearly, this would not be the case with a totally flat design.

A further natural benefit that arises out of design partitioning is easier design management. Sections of the design, partitioned into clusters of logically related functions, can be handed to small individual design teams or individual designers, or even outsourced to external third-party design consultancies.

Having argued the necessity and benefits of design partitioning and hierarchical design, the question then arises as to how many levels of hierarchy there should be in a design of a given size. How many blocks should there be at each level of the hierarchy? There are no definitive correct answers to these questions. However, some rule-of-thumb guidelines can be suggested. These are based on intuitive practices that tend to be used by the general design community.

Do not create unnecessary or artificial hierarchies or boundaries. Blocks should be grouped by functionality constrained by practical maximum sizes. As a guideline, the size of an individual block might lie somewhere in the 1,000- to 5,000-gate range. This is no longer a logic synthesis limitation. Rather, it becomes an issue of how understandable the block is to someone other than the designer. Of course, this guideline is heavily dependent on the logic content of the block. For example, a block implementing lots of basic registers can be gate-intensive, yet easy to understand, whereas a block with a complex state machine and complex control logic is going to become difficult to understand, once it reaches a certain size.

At any given level of hierarchy, try to limit the number of blocks at that level to six or seven. If there are any more, the schematics or hierarchy diagrams become cluttered, and it becomes more difficult to recognize the rationale behind the partitioning.

It is worth noting also that, if a bottom-up manual time-budgeted synthesis approach is used, each level of hierarchy detracts from the total available time budget. If an iterative characterize-compile approach is used, the number of iterations and the complexity of the synthesis scripts are likely to increase also.

A further incentive to avoid excessive use of hierarchy is to remember that, at the top of each level of hierarchy, we require a module that instantiates and interconnects all the modules at that level. This is a deceptively time-consuming exercise. However, the speed and reliability of this exercise can be improved dramatically if automated scripts are used to do this task. The efficiency of such scripts is greatly improved if interconnecting blocks use the same names for the same signals and port names match input and output signal names connecting to these ports.

4.2.10 Unused States in State Machines

Unused states in state machines are a source of potential problems. For example, if an encoded-state state machine designed around three state bits requires only five working states, this leaves three unused states. The problem arises when the state machine unintentionally enters an unused state. This can happen due to electrical noise interfering with normal operation or as a result of a poor design that does not manage asynchronous inputs properly. Either of these conditions could potentially send the state machine into one of the three unused states in the example above. If the CASE statement for generating next states does not list the unused state or simply lumps it into a default case, where the defined action is a "null" statement, the state machine may get stuck in the unknown state and stop functioning as intended.

There are several possible approaches to handling unused states. The most common is to lump unused states into a default CASE statement clause that puts the state machine back into the reset state on the next clock. This at least means that the state machine does not hang indefinitely. However, there can be side effects as a result of the interruption of the normal state sequences leading up to this event, and it may be difficult to detect what caused these.

If it is imperative that any branch into an unused state is detected in some way, an alternative approach is to send the state machine into an error-handling state following transition into an unused state or to generate an interrupt or assert a status bit when it happens. Software can then decide whether any remedial action is required.

Another approach sometimes used is to force the state machine to run through all the unused states once in sequence following a reset, then to loop around the normal used states, as required by the design. This technique ensures that the state machine cannot get stuck in an unused state. It has the further advantage in a design relying on functional test vectors, rather than scan-test, that these states then can be tested with normal functional vectors. If they were completely unused, there would be no way of exercising them with functional vectors.

4.2.11 Glitch Sensitivity

Glitch-sensitive signals should not be generated by decoding state bits. For example, it may be necessary to assert the *write* control pin on a memory cell for two or more consecutive different states of a state machine. If more than one state bit changes in any of the relevant state transitions, it is possible to get a glitch on the resulting control signal. The solution is to make the control signal a direct registered output itself.

A further more subtle glitch problem that can occur is the generation of unwanted glitches when switching a multiplexer to select one clock over another. Even though the circuit may be cleverly designed to ensure that the multiplexer changes state when both signals are at the same logic level, the output can still glitch. This arises as a result of different speed paths from the individual data and select inputs to the output. To avoid this particular problem, the ASIC vendor should be consulted for advice on selecting a glitch-free multiplexer type.

4.2.12 Modular Interfaces

Design time is reduced considerably if some up-front care is taken in defining consistent modular design interfaces. This is best illustrated by an example. Consider a design that has a memory access controller comprising two main functions. The first function is to select the highest-priority client request from N possible simultaneously requesting clients. The second function is to generate the required control sequence and data flows to allow the selected client to read from or write to the memory element.

If a number of the client blocks are designed by different designers before the memory controller block's interface is fully specified, it is almost inevitable that there will be several different protocols developed independently by each of the different designers. Even if the signal names are defined in advance, if their characteristics are not fully defined, each designer can make different assumptions. For example, one may assume that the acknowledge signal is a single cycle pulse while another assumes that the acknowledge signal is held asserted until the request is removed. This type of situation arises surprisingly frequently. Most commonly, it arises when a common service block like this is not assigned an owner initially and is not due to be implemented until later in the design cycle. In such a case, the interface signals are usually agreed on in naming terms and general functional terms, but all too often, some of the specific characteristics are not adequately tied down, such as interpretation of the acknowledge signal in the example above. In more extreme cases, if the common service block is not defined at all initially, other than in general functional terms but, for example, without interface details, the individual designers interfacing to such a block can go off and design completely different client interface protocols to the common service block.

The main problem with all of this is that, when a designer finally comes along to design the common service block, he or she will have to spend far more time designing unique interfaces to each client block. It is also more difficult to understand, test and debug multiple interfaces where one modular interface would have sufficed. Some up-front investment in clearly defining a common interface protocol avoids these problems.

The lesson is to define modular interfaces up front and in sufficient detail before any design work begins. If the project plan is created with this issue in mind, ensuring that the common service blocks are among the first to be designed, the potential problems or inefficiencies described above are unlikely to occur.

4.3 Coding for Synthesis

4.3.1 Inferred Latches

Take care not to infer latches where latches are not intended. This is a very common pitfall. It occurs when incomplete case statements or incomplete IF-ELSE statements are used. Incomplete case statements are not allowed in VHDL. However, they can exist in Verilog. This type of mistake is usually easily spotted in code reviews because experienced reviewers know to watch out for it. If the code is not adequately reviewed, the designer should always examine the synthesis report files because these provide an analysis of the number of flip-flops in the design. Any unexpected flip-flops reported in these files should be examined to see whether they were actually intended in the design.

As a general guideline, it is preferable to use flip-flops rather than latches when a storage element is required. There are, of course, situations where the specific required function can be achieved only by using a latch or where the gate-count overhead of using flip-flops instead of latches is critical. However, specification of synthesis constraints and static timing analysis operations are more complicated when latches are used, whereas, in general, using flip-flops simplifies these tasks.

4.3.2 Sensitivity Lists

Incomplete sensitivity lists constitute a further commonly encountered coding error. If a variable is omitted from a sensitivity list, the process will not be triggered in a behavioral simulation when this variable changes state. However, in gate-level simulations, the state change will cause the process or synthesized structure to evaluate. This can lead to differences between behavioral simulation and gate-level simulation results.

Incomplete sensitivity lists will often cause problems with the register-transfer-level (RTL) simulations, and they can take some time to identify. Doing a quick trial synthesis of the module before simulation starts is often useful because the synthesis tool gives a warning for incomplete sensitivity lists.

4.3.3 Registering Module Outputs

In a manual time-budgeted approach, it is easier to meet and specify timing constraints if all outputs are registered. This means that the block into which the outputs are feeding has almost a complete clock cycle to play with, and this is easy to specify in synthesis constraints. Contrast this with an approach that does not register all outputs and allows perhaps 50% of the time budget for register to output propagation via combinational logic in one block. The remaining 50% of the time budget allows for propagation of this output through a second block to another register input again via more combinational logic in the second block. It is difficult to predict in advance how many levels of logic may lie in the signal path in each block, and it is, therefore, easy to end up with one block overconstrained while another block completing the same logic path has excessive slack.

Registering outputs also makes a characterize-compile approach more effective, as it reduces the number of iterations required to balance the time budget fairly between two interconnected blocks.

4.3.4 Reset Values on Registered Signals

It is a good idea to put a reset value on all registered signals. This avoids the common problem that arises at the start of system gate-level simulations when, because of uninitialized register outputs, X-values are propagated throughout the design, and the simulation fails. It can take several days to find and fix the source of all X-propagation in a medium-sized chip. The problem is avoided by adding a reset value to all register structures. In some cases, for example, shift registers, it is not necessary to reset all the flip-flops. If the initial flip-flop in the chain takes on a known value at reset, the flip-flops further down the chain take on known values with each successive clock cycle as the initial known value is propagated through the chain. This approach works, provided that the output of the last flip-flop in the chain is not used or sampled before the known value has had a chance to ripple through. As a general rule, however, unless the design is gate-count-sensitive, it usually makes life easier to put a known reset value on all flip-flops. However, it should be noted that this can impact the maximum operating speed if the reset logic is implemented by placing a multiplexer on the input path to a flip-flop on a critical timing path.

4.3.5 State Machine Coding Styles

There are numerous coding styles for state machines. The most common is probably random encoding, where the designer does not particularly care what state assignments are used for the various states. However, there are some cases where it can be useful to use particular coding styles.

Gray code, where each successive state differs by a maximum of one bit from its neighboring states, can be used to avoid the possibility of entering unintended states when next-state decisions are based on asynchronous inputs. However, if there are multiple branches in the state machine, it is not always possible to make all transitions differ by only one bit without extending the number of bits in each state, which results in many redundant states. Gray-coding also has the advantage that it usually minimizes the next-state decoding logic. If the reason for using gray-coding is to manage branching safely, based on asynchronous inputs, it is only necessary to ensure that the states on either side of the branch decision differ by only one bit.

One-hot encoding is a coding style that asserts only one of the state bits for each possible state. This means that a large number of flip-flops is required, compared with encoded-state state machines. For example, a 32-state state machine requires only five flip-flops in an encoded-state implementation but 32 flip-flops in a one-hot implementation. However, the difference in resulting gate count may not be quite as extreme as this would suggest. One-hot encoding results in less decoding logic. Additionally, one-hot encoding may suit FPGA-type designs, where the number of flip-flops available is fixed and relatively high. The main reason for using one-hot encoding is to achieve a faster design.

It is also possible to implement a coding style that uses one or more of the state bits themselves as outputs. The state sequence is then planned in a way that generates the required output pattern on the selected state bit or bits. This method saves an additional flip-flop that would otherwise be used in addition to the state-bit flip-flops.

Whichever coding style is used, the state machine should be constructed so that its states are defined as constants or in the case of VHDL user-defined types, rather than hard-coded numbers. This approach significantly eases understanding of the code if the states are assigned names that give some indication of the function of the state. For example, a state transition from the state WAIT_START_PULSE to the state ACK_START_PULSE has much more instant meaning than a transition from the state *0101* to the state *1100*. Additionally, most timing waveform tools will display the state name, rather than the state number. This eases debugging considerably. Some synthesis tools can also automatically generate specified state-coding styles when the state machine is coded in a prescribed style and a synthesis directive is used.

4.3.6 Multiplexers and Priority

The coding style used to infer certain logic functions influences the architecture that the synthesis tool chooses to implement the function. For example, multiplexers can be inferred using either IF-ELSE or CASE constructs. An IF-ELSE statement can have inferred priority, in contrast to a CASE statement that could implement the same function. The interested reader is referred to the synthesis tool documentation to learn more about this topic, because some of the details are quite subtle. The method is sometimes used deliberately to ensure a short timing path on a late-arriving input. It is useful to note that an IF-ELSIF construct results in the same logic as a CASE statement when the IF-ELSIF construct covers all possible branches. However, for complex branching, it is advisable to use CASE statements in preference, unless a priority is specifically required. Otherwise, unnecessary and unwanted priority structures that result in larger gate counts may creep into the design.

4.3.7 Resource Sharing

Resource sharing refers to the sharing of physical structures (e.g., adders, comparators, etc.) to perform operations on inputs that are executed at mutually exclusive times. As an illustrative example, consider Figure 4-1. This shows a variable Z gener-

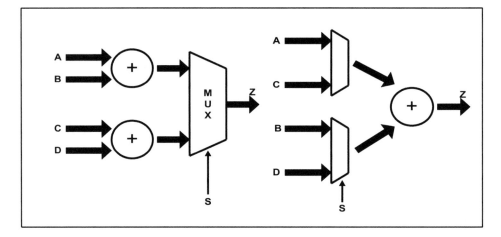

Figure 4-1   Resource-Sharing Example

ated by adding either two input variables *A* and *B* or two other input variables *C* and *D*, with the choice of inputs being determined by a control signal *S*.

Consider the following two possible ways that this operation could be physically structured. One approach is to have two separate adders with a multiplexer routing one of the two results to *Z*. An alternative approach that resource shares the adder is to have multiplexers on the inputs to a single adder.

When coding in Verilog, Synopsys Design Compiler will resource share the adder if the addition operations are specified in mutually exclusive branches of an IF-THEN-ELSE statement, but will not resource share the adder if the addition operations are specified in a single statement using the Verilog condition operator (?).

The following Verilog code results in the adder being shared.

```
if (S)
    Z  =  A+B;
else
    Z  =  C+D;
```

However, the following code that uses the conditional operator results in separate adders being used.

```
Z  =  (S) ? (A + B) : (C + D);
```

The reader is referred to the synthesis tool user manuals for a fuller treatment on the topic of resource sharing.

4.3.8 Constructs That Will Not Synthesize

It is important to use only synthesizable constructs in the chip code. For example, WAIT statements cannot be synthesized. The synthesis tool similarly ignores delays in assignment statements, so do not write chip code that relies on such delays to function properly. These types of constructs are intended for testbenches and certain types of models. However, both testbenches and models should ideally avoid using these constructs also, or at least try to confine their usage to a small number of modules. The reason for this is that two methods of speeding up verification (hardware acceleration and cycle-based simulation) are both more effective if as much of the code is synthesizable as possible.

4.4 Coding for Testability

Under this heading, we can consider coding for improved functional simulation and functional verification coverage, as well as coding issues related to manufacturing test vector coverage. In fact, these topics overlap when functional test vectors are used as the method of providing manufacturing test vectors.

4.4.1 Coding for Functional Simulation/Functional Verification

There are numerous miscellaneous tips that could be given under this heading. A few examples are covered here.

4.4.1.1 Observability of Internal Signals via External Pins

Remember that, unlike the case with RTL simulations, you cannot automatically observe internal signals when testing a chip on the bench. Indeed, even with gate-level simulations, some of the observability disappears, or at the very least, ease of observability is reduced. One approach that attempts to redress this situation is to bring sets of key internal signals out to external pins. There is usually no need to use dedicated I/O pins for this purpose. Instead, select a handful of noncritical pins that have no active functional purpose for many of the tests. For example, there may be a parallel data interface coming out of the chip that is used only in one particular mode. Having selected the pins, internal register settings can be used to put the chip into test mode. In the test mode, key internal signals are multiplexed out to the selected pins instead of the normal functional signals that would drive these pins. A small number of selected output pins, along with a few register-selectable multiplexer settings, can be used in this way to observe many internal signals. For example, eight output pins, along with three register bits used as the select inputs on a multiplexer, allows up to 64 internal signals to be seen, i.e., eight sets of eight signals.

Among the useful sets of signals that one might like to observe in this way are state-variable bits from complex state machines, selected status bits, selected clocks, internal counters, etc.

4.4.1.2 Observability of Internal Signals via CPU-Accessible Registers

The same principle can be applied to observing internal signals via CPU-accessible registers. However, in this case, instead of routing the signal buses via multiplexers to external pins, they are instead routed via multiplexers to dedicated CPU-addressable registers. The disadvantage of this approach over the external pin approach is the reduced rate of updates or snapshots of internal buses. A CPU read of a register can take multiple cycles of the ASIC internal clock. During these cycles, a state machine may have changed state several times, so, in effect, we see only subsamples of the changes. However, the technique is quite effective for observing relatively slow-changing status signals or for finding out what state a state machine locks in should it enter a lock condition. Debugging the effects of a locked state machine problem can be extremely difficult. Knowing the state in which the state machine is locked gives a significant starting advantage.

4.4.1.3 Observing Internal VHDL Signals

Unlike Verilog simulations, where internal signals can be seen at the testbench level in a simulation by referencing them with full hierarchical path, this cannot be done in VHDL. However, there is a method of working around this. If the signals of interest are declared in a global package, which is declared in the internal VHDL module, the same signals can also be made visible at the testbench level by declaring the package there, also. This eliminates the need for propagating the signals through the levels of hierarchy.

4.4.1.4 Modeling Large Memory Elements

Many designs rely in some way on the use of memory elements. If these are small or infrequently accessed, the method of modeling them may not be critical. However, if they are large and frequently accessed, the modeling method can have a significant impact on simulation storage requirements and speed.

The memory contents can be modeled in many ways. We will briefly examine two in the VHDL context (integer versus *std_ulogic*), just as an introduction to the topic. If each bit in memory is modeled as a *std_ulogic* entity, depending on implementation, the simulator may require a nibble to store each *std_ulogic* bit. Clearly, for large memories, the overall storage required to model the memory could become very large and possibly impractical. If it becomes sufficiently large that page-swapping is required, this, of course, has an influence on speed, as well as storage requirements. If the word width is sufficiently wide, it may be far more efficient from the point of view of storage to model the memory contents as an array of integers.

However, there is a certain execution time lost in converting the integers to and from *std_ulogic* format when accessing the memory, assuming that the design is largely based on *std_ulogic* types. Additionally, it is not as easy to represent *X* or *Z* types in an integer memory model. Special dedicated values would have to be used and conversions carried out at the memory model interface.

As a general guideline, it is probably better to use integers to represent large memories using wide word widths. However, if the decision is likely to be a critical one, the topic deserves considerable analysis by the design team.

4.4.1.5 Absolute Pathnames for Input or Output Files

Testbenches or "signal monitors" within the ASIC code itself often use external files to source input data or store output data. The code should not use absolute pathnames for these files, because the pathnames will no longer be valid if the design is transferred to a different directory level or the same directory level with different upper-level paths. Instead, the files should use relative pathnames where each file location is specified relative to a design root directory that will always exist, no matter where the design tree is reproduced.

4.4.1.6 Read-Writable Registers

Try to avoid coding write-only registers. By making them read-writable, the simple test of reading back the written value tests quite an amount of logic. This is an important functional behavioral test and a useful method for generating test vectors if a functional test vector approach is being used.

4.4.1.7 Preloadable Counters

At the expense of a modest increase in gate count, it can be very useful to be able to preload counters to any desired preset value via the CPU interface. This can speed up simulation of certain types of behavior by multiple orders of magnitude. For example, let's assume that a certain change of behavior is expected from a design after 1,023 maximum-length data packets have been received, e.g., a specific status flag is asserted. To simulate this may take from several hours to several days, depending on processing power and complexity of the design. The same result may be verified by presetting the packet counter to 1,022 and simply sending in one maximum-length packet. This simulation will probably execute 1,000 times faster. Another common use of this technique is to simulate overflow conditions in addressable FIFOs. The read and write address generators for a FIFO are often based on some form of counter. To speed up simulation of the response to an overflow condition, the addresses can be preloaded to chosen values that precipitate the occurrence of the overflow condition.

4.4.1.8 Split Counters for Functional Test Vectors

To gain maximum toggle coverage with the minimum number of vectors in circuits using large counters, the counter can be split into smaller sections using a test mode. To illustrate the point, consider a 12-bit binary counter. To ensure that all the output bits toggle requires a minimum of 2 K clock cycles—the most significant bit will have toggled at this stage. If in test mode, the counter is split into two 6-bit halves; it only takes 32 or 33 cycles to ensure that all bits have toggled.

4.4.2 Scan-Test Issues

A number of common issues arise in scan-testable designs.

4.4.2.1 Multiple Clocks

Each individual clock should be controllable from an external pin. One approach is to have each internal clock source multiplexed with an externally driven clock. The external clock can then be used to clock all flip-flops in an externally selected scan-test mode.

It will also make life easier if the flip-flops in each individual clock domain form their own scan chains, rather than mixing flip-flops from different clock domains in the same scan-chain. The reason for this is that, although skew within a given clock domain is usually well controlled, there will be a bigger variation in skew when a single scan clock drives a scan chain spanning multiple clock domains.

If flip-flops in different clock domains must be connected in the same chain, perhaps due to an upper limit on the allowable number of chains, there are still some steps that can be taken to help resolve potential timing difficulties.

Form a subchain for each clock domain, then interconnect the subchains, rather than randomly intermixing flip-flops from each domain into the one chain.

If the source clocks for each subchain can be independently controlled, create a deliberate small phase shift between the clocks. Use the earliest clock to clock the subchain at the end of the total chain first, and work progressively backward. For example, if the total chain consists of two subchains, use the early clock to clock the second chain in the subchain and the later clock to clock the first chain. This reduces the possibility of having a hold time problem that could occur at the interconnection point of the two chains, arising out of the fact that the two chains' clock networks are skew-balanced only within their own network and not skew-balanced with respect to each other. The situation can be further improved by adding a buffer between the end of the first chain in this example and the beginning of the second.

4.4.2.2 Internally Generated Flip-Flop Resets

If flip-flops are reset by internal functional logic, the random pattern inserted during scan testing may throw the flip-flop unintentionally into a reset or preset state. This flip-flop cannot, therefore, become part of the scan chain because going into a reset or preset state interrupts and corrupts the transfer of scan data. Consequently, the flip-flop itself and, more than likely, some of the closely associated random logic cannot be fully tested. A workaround for this is to multiplex the reset or preset signal with a test-mode input that ensures that the flip-flop cannot be reset or preset when in scan test mode.

4.4.2.3 Tri-State Buses

If multiple tri-state drivers drive an internal bus in the chip, there is a possibility that more than one driver can be enabled during a scan test sequence, thus causing bus contention. To avoid this, the enable signals for the tristate drivers should be multiplexed with enable signals that are driven from a decoder that allows only one driver to be enabled at any instant. In scan-test mode, the multiplexer is forced to select the decoder outputs to drive the tristate enables instead of the normal functional sources. This ensures that only one driver is ever enabled at a time and avoids the contention problem.

4.4.2.4 Bypassing Memory Elements

If a memory element cannot be scanned, as is often the case, the input data, control and address lines become unobservable, and, similarly, the output data, status lines and any following logic cannot be controlled. This can result in a significant drop in test coverage. A simple way around this is to multiplex inputs through to outputs in test mode. This increases coverage significantly. To test the memory elements functionally themselves, the inputs should be routed to input pins of the ASIC, where they can be driven directly in specific test modes. Alternatively, use built-in self-test (BIST) modules, which are available for some ASIC vendors' memories.

4.5 Coding for Multiple Clock Domains

4.5.1 Use of Multiple Clocks

It is usually easier said than done but, if possible, it is generally a good idea to limit the number of different clock domains in a chip. Multiple clock domains can complicate synthesis and scan testing. They also increase the possibility of encountering metastability problems unless special care is taken with signals crossing clock boundaries. If a number of signals cross from one clock domain to another, it is a good idea to have a special dedicated block or blocks where all boundary crossing takes place. This means that all other blocks can be synthesized more easily because they do not have to deal with multiple clocks. The necessary synthesis techniques for managing the boundary crossing signals can be concentrated in one or several specific modules dedicated to boundary crossing signals. This also makes the design easier to review, because reviewers need watch out only for boundary crossing problems in a specific small number of easily identified modules.

4.5.2 Crossing Clock Boundaries

4.5.2.1 Signals Crossing Clock Boundaries

The most common problem with signals crossing clock boundaries is that of metastability. If the clocks do not have a fixed phase relationship, the possibility of this happening may be quite high. Metastability can occur when a flip-flop input signal is changing state at or near the instant that the active clock edge is occurring. Under such conditions, the output state after the clock edge may be indeterminate or may oscillate, and it takes a finite amount of time for the flip-flop finally to settle at a stable high or low state. This can cause multiple problems. For example, it can send a state machine into an unknown or unused state. Consider the simple state machine example in Figure 4-2.

In the simple state machine of Figure 4-2, there are two flip-flops and three states: 00, 01, and 10. B represents an input synchronized to *Clock-B* (for example, a 25-MHz clock), while the state machine runs off *Clock-A* (for example, 30 MHz). If B transitions on or near an active edge of *Clock-A*, there is a possibility that one or the other or both of the state flip-flops will briefly enter a metastable state. Consider a scenario where the delay between B and the most significant state flip-flop is slightly less than the delay between B and the least significant state flip-flop. In such a case, the most significant flip-flop may see the transition from 0 to 1 on B just before the active edge of *Clock-A*, and, therefore, this flip-flop will transition from 0 to 1 in response, whereas the least significant flip-flop may not have seen the transition in time and, hence, becomes metastable, finally resolving at logic 1. This scenario puts the state machine into an undefined "11" state, in which it becomes stuck.

A simple solution to the above problem is to partially gray-code the state machine as shown in Figure 4-3. A gray-code sequence is characterized by the fact that, in going from one code in the sequence to the next, there is never more than 1 bit in the code that changes state. If the transition involving a decision based on the value of an asynchronous input (such as B in the state machine below) is in a gray-coded section of the state machine, the state machine will not depart from its planned sequence, even if metastability occurs. In the example of Figure 4-3, the transition from the second to the third

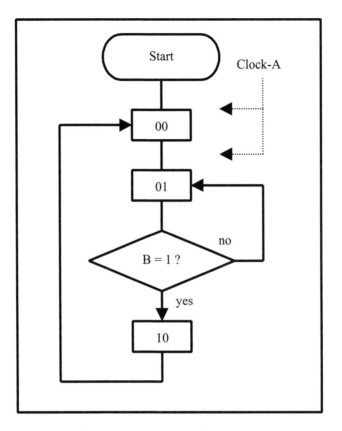

Figure 4-2   Simple State Machine (Random State Coding)

state involves only one change of state bit. The least significant state bit does not change based on the decision and, therefore, it does not matter whether it sees the transition in time or not. The most significant state bit either sees it in time, in which case we move from state two to state three, or, if it does not see it in time, it stays in state two. In both the above cases, the state machine follows its intended sequence. The only impact is that of a possible latency of one clock cycle in reaching the final state.

An alternative solution to managing boundary crossing inputs is to synchronize the input signal to the timing domain of the circuitry which is processing it. In terms of Figure 4-2 or Figure 4-3, this involves registering the asynchronous input from the *Clock-B* domain into a synchronizing flip-flop in the *Clock-A* domain, as shown in Figure 4-4. Note that the duration of signal *B* must exceed the period of *Clock-A* for this technique to work.

The idea here is that the output signal (*sync-B*) of this circuit is used to drive the

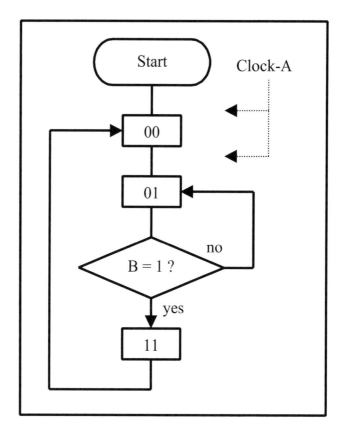

Figure 4-3   Simple State Machine (Partially Gray-Coded)

decision in the state machine of Figure 4-2 instead of the raw unsynchronized signal *B* itself. While the second flip-flop in Figure 4-4 may briefly enter a metastable state, the idea is that if clock period *A* is sufficiently long relative to the metastability resolution time of the flip-flop technology used, the output *sync-B* will resolve to a stable state sufficiently early in the periodic cycle of *Clock-A* that it is guaranteed to be stable as an input into the state-machine clocked by *Clock-A*, i.e., it has become synchronized to the *Clock-A* domain.

Note that with the synchronizing flip-flop approach, there is effectively an extra clock cycle added to the delay on the path from input *B* into the state machine. The application must be such that this is acceptable if this type of solution is chosen. If the delay is not acceptable, a gray-coding solution may be a better approach.

4.5.2.2 Buses Crossing Clock Boundaries

A relatively common mistake that is made by inexperienced designers is to try to

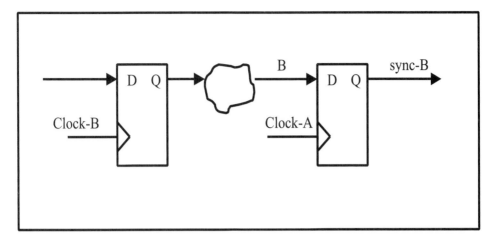

Figure 4-4 Synchronizing an Input Signal

apply a simple synchronizing flip-flop approach to synchronizing bused signals. For example, let's say that we have a byte-wide input data bus *B[7:0]* from the *Clock-B* domain. It is not sufficient to pass each individual bus element through its own synchronizing flip-flop to create the synchronous bus *sync-B[7:0]*. Although such an approach ensures that no individual signal in this bus remains indeterminate shortly after the active edge transition of *Clock-A*, there is no guarantee that the bus data alignment is preserved; i.e., due to clock skew, some of the synchronizing flip-flops may have seen the transitions on their inputs in time, while others have not. This means that the output of the synchronizing flip-flops may be a mixture of *B* data values from two consecutive sampling instants.

A possible solution to synchronizing the bus is to have a *data_valid* signal accompany the *B* bus. Provided that the *data_valid* signal is not asserted until a comfortable timing margin (e.g., half a clock cycle) after all the *B*-bus bits have transitioned and provided that none of the bus bits transition again until a comfortable margin after *data_valid* has deasserted, it is sufficient to synchronize the *data_valid* signal only and use this signal to indicate that the *B*-bus bits are stable and that they can then be used directly (e.g., registered safely) in the *Clock-A* domain. A further requirement for this solution to work is that the *data_valid* signal must have a minimum assertion period greater than the period of *Clock-A*. This solution is practical only if the rate of updates on the *B*-bus is much lower than the rate of resampling in the *Clock-A* domain.

An alternative solution to managing bused data across an asynchronous boundary is to use a FIFO that has separate read and write clocks. In our example, the successive samples of bus *B[7:0]* are clocked in to the FIFO on the *Clock-B* and read out on the

Clock-A as they become available. In such a solution, it is quite common to have an address counter running off each of the clocks to check that the write-clock does not catch up with the read-clock and, consequently, overwrite unread data and, conversely, to check that the read-clock does not catch up with the write-clock and read out old or uninitialized data. This is another common source for error. If these two address counters are compared to determine their distance apart, the counters must be gray-coded. If they are not, because they are each asynchronous with respect to the other's clock domains, the comparison may yield false results. The only safe way to make the comparison is to gray-code the counters.

4.6 Summary

This chapter has covered a wide range of issues under the heading of design tips and guidelines. Beginning with some general common-sense guidelines on coding style that assist in making the code more easily understood and reviewed (and consequently, less error-prone), it then progressed onto a wide range of specific tips and guidelines that address issues arising in many companies across many and varied design types. Common synthesis and design for test topics were covered, and the frequently problematic subject of dealing with multiple clock designs was addressed. The reader is advised to reread this chapter at the start of each new design project. It is a compact source of tips for novices, and even seasoned designers may be usefully reminded of one or two issues that they do not always automatically remember.

CHAPTER 5

ASIC Simulation and Testbenches

5.1 Introduction

Simulation is the process that tries to prove that the ASIC will perform the correct functions or algorithms within the constraints of the target system. As ASICs continue to grow in size, designing the top-level testbench becomes almost as challenging as designing the target ASIC itself. The top-level testbench must be available before any chip-level simulations can start. The probability of bugs passing through undetected to the final silicon is significantly reduced if the testbench is designed using a quality approach. For these reasons, testbench design and planning should be treated as mainstream activities and not just rushed together hurriedly as the top-level integration date approaches. Depending on the size and complexity of the ASIC and its interfaces, one or more engineers should be assigned the dedicated role of testbench engineers (see Chapter 12, "The Team").

To get maximum value from a well-designed quality testbench, a comprehensive list of simulation tests is required. Although there are typically only one or two engineers tasked with designing and implementing the testbench, the wider design team can have a role to play in defining and running the tests that use this testbench. However, this introduces a philosophical debating point. One school of thought is that designers will specify only tests that test their interpretation and implementation of the design specification. This school of thought advocates that it is, therefore, better that the tests are specified by a dedicated test team not involved in the design. Such a team may interpret the design specification in a different way, and this acts as a safeguard or a

93

cross-check on interpretation of the specifications. A second school of thought argues that only the design team members know the characteristics and intricacies of the design sufficiently well to be able to specify a sufficiently comprehensive set of tests to exercise it. A possible compromise view, resources permitting, is to have an independent test team specify the test specification, then to have this thoroughly reviewed by the design team to ensure adequate test coverage. Regardless of which approach is taken, creating a test list document early in the project will help ensure that the testing is done in a controlled manner and that the design of the testbench will encompass all the features required to execute the complete set of tests.

This chapter looks at some of the issues involved in testbench design and ASIC simulation. The component parts of a testbench are explained. Possible simulation strategies are considered, and methods of speeding up simulation times are suggested. The different types of simulation tests are distinguished and explained. The chapter finishes by introducing some of the issues associated with capturing manufacturing test vectors, using the system testbench.

5.2 Quality Testbenches

A well-designed testbench that is ready on time will help ensure that the project runs according to plan. Because the testbench can be the one of the most complex parts of the system, its constituent modules should be designed and architected using similar techniques to those used for designing the ASIC modules themselves. The components of a typical testbench are shown in the simulation environment of Figure 5-1. A unit under test (UUT) is exercised by testbench drivers that generate input stimuli to the UUT. Monitors then capture outputs from this UUT. Sometimes drivers and monitors are merged into combined modules. Input files can be used to control the drivers and output files used to store vectors captured by the testbench monitors. The UUT can be ASIC modules or an entire chip. The drivers are sometimes referred to as *transactors* or *bus-functional models* (BFMs), depending on the style of implementation.

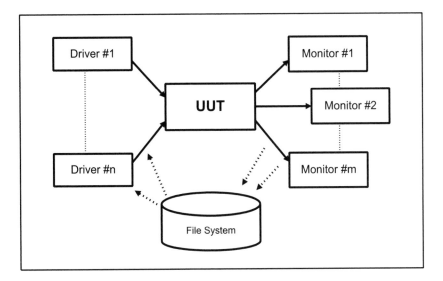

Figure 5-1 Typical Testbench Components

A quality testbench should have as many of the following characteristics as possible:

- It should be debugged and tested itself before being used to test UUTs.
- It should be configurable.
- It should be relatively easy to use.
- It should be capable of generating a wide range of input stimuli.
- It should be capable of logging monitor results to output files.
- It should facilitate easy insertion of additional drivers and monitors.
- It should be self-checking.
- It should facilitate the generation and capture of manufacturing test vectors.

Simulations are required at a number of different levels within the design. Individual module testing, subsystem testing, chip-level testing and board-level testing typically all require unique or custom testbenches. However, with careful planning, it may be possible to reuse some of the same drivers and monitors at different levels, albeit in different overall testbench architectures.

5.2.1 Generating Input Stimuli

Input signals/stimuli generally fall into the categories of address, data and control lines. The testbench and its associated drivers must support generation of a wide enough combination of these signal types to test each function of the module or chip under test. A good driver module will support flexible generation of input signals at a relatively high level of abstraction and avoid forcing the user to specify individual signals in low-level detail. For example, a driver that models a packet generator for a data-comms chip might support a relatively high-level command interface with the following command format:

```
send_pkt(source_addr, dest_addr, nr_bytes, pattern_type,
    error_flag, error_type);
```

where

- *source_addr* ...specifies the port address of the data source
- *dest_addr*...specifies the port address of the destination port
- *nr_bytes*...specifies the number of bytes in the packet
- *pattern_type*...selects the data pattern
- *error_flag*...indicates whether or not the packet contains some type of error
- *error_type*...identifies the error type, when *error_flag* indicates that the packet should deliberately contain errors

Note the inclusion of error fields in the above example. For many drivers, it is important to have the ability to generate error sequences to ensure that the system being tested responds appropriately.

For transactors of this type that model input data streams, it is useful to be able to select different patterns of data easily:

- Incrementing data streams are useful for debugging . It is easy to spot the data sequence as it travels through the internal data paths of the chip.
- Random input data can trigger unexpected data-dependent errors that might otherwise be difficult to trigger.
- Selecting data patterns from presetup reference files can be used to compare captured outputs with golden or reference output files to verify that the chip is processing the input stream correctly, as defined by the operating algorithms.

One consideration to take into account when modeling data streams is whether or not the driver should be synthesizable. If data is generated internally, rather than read in via text I/O, it has the advantage that the driver can potentially be fully synthesizable.

This is useful for cycle-based simulators and for certain hardware acceleration techniques.

In addition to being able to generate random input data patterns, it is sometimes useful if the driver or transactor can model random latencies and delays. Continuing the previous example of a datacomms chip, it may be useful to be able to trigger several packet generator drivers to start transmitting at intervals separated by a random (or, more precisely, pseudo-random), number of clock cycles because this may more accurately represent real-life applications. Another example where random input timing is useful is in modeling asynchronous inputs, where input transitions can be set up to occur randomly between two successive clock transitions.

5.2.2 Running the Tests

It is important that tests can be configured and run quickly and simply. The use of makefiles for compilation scripts ensures that only the testbench or ASIC code that changed is recompiled when modifications are made. For large chips, this is much faster than recompiling the entire testbench. The use of properly structured configuration files facilitates setting up new tests quickly and easily. For regression testing, it is desirable that all existing tests can be easily reproduced and re-executed. (Regression testing refers to the procedure of rerunning a subset of the tests following modifications to ensure that the modifications have not inadvertently caused unexpected problems in previously working sections of the design.) Again, this is an easier exercise when well-structured configuration files are used. Existing tests can be rerun in batch mode using scripts simply by copying in new configuration files and/or driver input files between tests.

As an alternative to configuration files to control simulations, it is possible to use simulator features that allow C modules to be integrated into the testbench. These modules can utilize all C language constructs and can generate sophisticated graphical user interfaces to control the testbench. However, it is important that the testbench can be run automatically to allow easy regression testing.

5.2.2.1 Testbench Configuration Files

For large system simulations, it is advantageous not to have to recompile the testbench for each new test. This can be relatively easily achieved through the use of test configuration files. Test execution begins by parsing the configuration files for directives on how to configure the simulation. To exercise or test different functions, the configuration file settings and input data file selections need to be changed but there is no requirement for recompilation.

A master configuration file can specify which drivers and which monitors are enabled. Control over which nodes and which functions are exercised in a given test (i.e., what the test does) is usually determined by two factors:

- The input files for the various testbench drivers govern the sequence of input test vectors applied to the ASIC.
- For ASICs with internal programmable registers, the values programmed into these registers determine the operating mode and characteristics of the ASIC for a given test.

5.2.2.2 Input Files for Testbench Drivers

There are numerous possible ways of specifying the input files for the testbench drivers. In the simplest style, each driver always reads its input data from a fixed file name. For example, an RS232 driver could always read its input data from a file called *rs232_driver_input.txt*. If this fixed file style is used, it is important that the file name be specified in the driver code using relative, rather than absolute, directory paths, so that the testbench is portable. In a testbench employing fixed input file drivers, the contents of the fixed file must be changed for different tests. To reproduce tests using this type of driver, each variation of the input file can be stored in an input file directory with its own unique name, e.g., *rs232_test1_input.txt* , *rs232_test2_input.txt*, etc. Rerunning a specific test then involves copying (or making a link to) the appropriate file to the location where the driver expects to find it, and renaming it with the generic name of *rs232_driver_input.txt*. Alternatively, each test can have its own directory structure, where the unique copy of the fixed input file name differs in content from directory to directory and, therefore, specifies different test characteristics.

Another approach to specifying input files for the testbench drivers is to specify the input file name for each driver in the master configuration file. With this approach, each test requires its own unique copy of the master configuration file. This is most easily achieved by having a separate directory structure per test. A sample extract from a master configuration file that specifies the driver input files might look something like this:

```
# enable the RS232 driver
RS232_DRIVER  : ON
RS232_DRIVER_IP_FILE : rs232_test4_input.txt
# disable the parallel port driver
PARALLEL_PORT_DRIVER : OFF
# parallel driver input file irrelevant as driver is off.
PARALLEL_PORT_DRIVER_IP_FILE  : par_port_test3_input.txt
```

5.2.2.3 Operating Modes

For ASICs with internal programmable registers, the operational mode of the ASIC for a specific test can be set by programming the ASIC's internal registers to a specific set of values. For example, an MPEG encoder/decoder chip may require different register settings for its two main operating modes (encoding and decoding). To test a specific mode, therefore, requires programming various internal ASIC registers to mode-specific values.

Programming internal ASIC registers in the simulation environment is usually done by using a CPU driver. Depending on how simple or complex a CPU driver is, many different formats are possible for the associated input data file. A common CPU driver approach is to support a small set of simple pseudocode instructions, most of which are ASIC register read/write instructions. A typical simple CPU driver input file might look something like this:

```
WAIT 100;      -- wait 100 cycles
WRITE  0001 FC00;  -- enable interrupts
READ FC10;     -- poll status register
```

Of course, a better driver will use predefined constants with meaningful names for the addresses, because this makes the sequence of events more understandable:

```
WAIT 100;      -- wait 100 cycles
WRITE  0001 INTERRUPT_REG;  -- enable interrupts
READ STATUS_REG;    -- poll the status register
```

When designing the CPU driver, it is worth considering the format of the input file carefully. It may be possible to use the same file format for hardware testing; this allows hardware and simulation testing to share the same CPU test files. Even if the format is not identical, some forethought should ensure that the formats are sufficiently similar that it is easy to translate automatically from one format to the other, using relatively simple scripts. This facilitates reproducing tests from one environment in another environment.

5.2.3 Comparing and Logging the Results

Testbench monitors can log the activity that occurs on various buses in the design. These can include internal buses, buses at the I/O boundaries or external buses. Internal buses, however, may disappear or at least be renamed after synthesis, so it advisable not to rely too heavily on monitoring internal buses. The extract below represents a typical format that might be used in a log file.

TIME	ADDRESS	DATA	R/W
.....	.....	.....	.....
2100	8000	FF	R
2133	8004	EF	R
2199	8004	00	W
.....	.....	.....	.....

A common method of building a self-checking testbench is to compare the output log files captured during a specific test with expected values from reference or "golden" log files generated for that test. If the files match as required, the test is deemed to have passed.

Self-checking test suites are more or less essential for chips of any complexity. Determining whether tests pass or fail by examining waveforms on a waveform viewer is not only error-prone but also incredibly slow. Such a manual form of testing does not lend itself to regression testing—it is simply too slow. In a self-checking testbench, output file comparisons can provide the pass/fail indication, and, in the case of failure, the precise time of failure can be identified, provided that there is a simulation-time column in the log files.

It is important to be able to enable and disable testbench monitors selectively because they can generate very large data capture files. For example, if there are 10 monitors in a complex simulation, and only five of them are relevant to a particular test, there is no point enabling all 10 of them. The extra five will generate unwanted data files and can slow down the simulations considerably. The on/off status for each of the individual monitors can be set up in the central master configuration file, as in the example below.

```
# enable the RS232 monitor
RS232_MONITOR   : ON
# enable the PCI-bus monitor
PCI_MONITOR   :   ON
# disable the parallel port driver
PARALLEL_PORT_DRIVER : OFF
```

In a more advanced testbench, it is possible to have more elaborate monitor controls than simple on/off settings. The monitors may accept input parameters (specified in the master configuration file) that control what types of transactions are recorded. For example, a simple on/off type memory bus monitor may capture the memory's address, data and control lines for all new transactions. A more elaborate memory bus monitor may be programmed to capture only a subset of all the transactions, e.g., read cycles only, write cycles only or accesses to a specific address range.

Sometimes, it is easier to interpret the results of a simulation by plotting the results on a graph. If this is a requirement, the relevant monitors should generate the plot data in plot-friendly format or at least in a format that requires minimal editing to allow it to be read by a plotting package. Some simulators have built-in plotting capabilities.

5.3 Simulation Strategy

ASICs typically connect to a CPU, sometimes directly or other times via a system bus, such as a PCI bus. The product test programs and application code interact with the ASIC in the integrated target product. One approach to testing this type of ASIC is to apply some of the application or test code in the simulation environment and involve software engineers in the ASIC simulation phase. This approach has a number of advantages:

- The software group becomes part of the team, fostering good cooperation between hardware and software engineers. This close cooperation can be useful at a later stage if, for example, there are problems with the silicon that a software workaround can fix. This allows the product to be launched without the need to respin the ASIC.
- Software groups normally have a very good quality approach to techniques such as source code control, fault tracking, documentation, etc. These techniques can be learned and used by the hardware group.
- The software engineers will test the ASIC in the manner that it will be used in the real application, using the register accesses defined in the specification.
- The software engineers can build up a library of low-level software driver function calls that configure the ASIC in specific operational modes. Each such function call may involve a reasonably complex sequence of register programming steps to put the chip into the required mode. An ASIC engineer familiar with only a small subset of the registers in the chip can then easily set up specific test modes, using the software provided library functions without needing to know details of parts of the chip for which he or she has no test or design responsibility.
- The software engineers benefit because their lowest-level driver code will be tested before the silicon becomes available. This will significantly speed up the silicon evaluation and hardware/software integration phases. Delays

often occur during these phases because the register specification is incomplete or inaccurate. This is difficult to diagnose with the real silicon, but internal registers are easily observed during simulations.

5.3.1 Software Hierarchy

Figure 5-2 shows a view of the classic hierarchical software model that is typically used in organizations writing embedded code and application code to run on ASIC-based hardware subsystems.

The software is broken into a number of different levels of hierarchical layers so that it is easier to manage and maintain, and easier to port to new hardware platforms. The different levels allow software engineers to concentrate on their particular focus areas without needing to understand all the levels in great detail. The higher the level of software, the more the code is abstracted from the hardware. The lowest level, called the *driver level*, accesses and configures the microprocessor, ASIC registers and RAM. The driver level contains the low-level functions that are needed to configure the ASIC and to control and monitor its operation. These functions must access ASIC registers via a CPU interface. Certain system operations may require that low-level detailed configuration data is programmed into a series of ASIC registers in a specific complex sequence. No layer of software other than the low-level driver layer should need to access ASIC registers directly. Each layer only communicates only with the layer immediately above and below it.

5.3.2 The Software Driver Testbench

Assuming a C-based software development environment, the low-level software driver code can be used in the ASIC testbench (see Figure 5-3). This driver code should contain all the functions needed to control and test the ASIC fully. There are three different methods in which the driver code can be incorporated into the testbench. The first option is to use a C model cosimulation feature, which is available in most HDL simulators. The second is to use a cosimulation tool. The third is to write a program to convert C into Verilog automatically. Provided that the code used for writing the low-level drivers does not use any complex C structures, the conversion from C into Verilog should be relatively simple.

The top-level testbench will typically encompass the ASIC and the driver-level software module, and will also include testbench drivers that generate input data and compare/log output data. The driver-level module should execute a pseudo-program that configures the ASIC for a particular function, and typically performs some operations during the processing of data. The top-level testbench can sequence the opera-

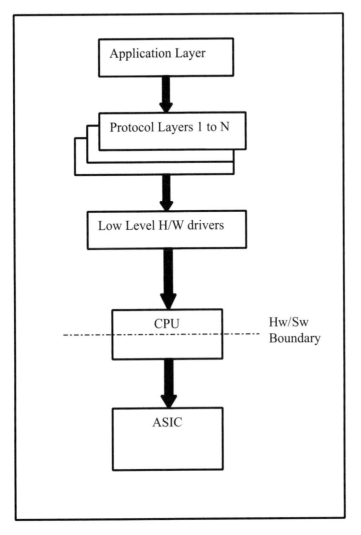

Figure 5-2 Hierarchical Software Model

tions of the microprocessor and the input data drivers.

5.3.3 C Model Cosimulation

This technique uses simulator features that are specific for each vendor tool. It typically involves creating a wrapper around the C code in VHDL/Verilog. The C code inside the wrapper will emulate a simple version of the microprocessor without modeling the detailed, complex structure of the processor. This simple model allows the sim-

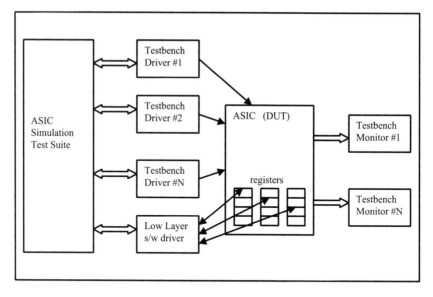

Figure 5-3 Software Driver Testbench

ulation to run faster without significantly reducing the quality of the simulation. The driver code will have access to UNIX/PC files and will often implement accesses to UNIX files quicker than using VHDL/Verilog text I/O functions. The lowest level of driver code will often need to call specific C functions to drive signals in the VHDL/ Verilog environment.

5.3.4 Co-Verification Tool

This technique is useful for testing some of the application and driver code in advance of having silicon prototypes. It simultaneously tests the ASIC and the application software in an integrated test environment. In a typical coverification environment, a synchronized communications interface exists between the software debug environment and the VHDL/Verilog ASIC simulation environment. Debugging the software/ hardware interface in this manner speeds up the development cycle considerably. Significant chunks of the software that traditionally were debugged following availability of prototypes can now be debugged in parallel with the ASIC simulation phase, in advance of receiving any prototypes. There is also the added benefit that certain software tests may reveal ASIC bugs that would not have been triggered in the traditional hardware-only simulation environment.

Coverification techniques and tools are rising in prominence as a result of the requirements of large SoC designs, where there is a move toward the use of system-level design languages and higher-level modeling of many of the component parts of the system. In such a design environment, different parts of the system are often modeled at different levels of abstraction, using a range of different languages. A coverification tool is a necessity in this type of design environment.

5.3.5 Converting the C Code

The lowest-level software driver is typically executing simple functions that do not need higher-level C constructs. Because C code is quite similar to Verilog, the code can be automatically translated, provided that it is written in a simple style with an agreed set of conversion rules defined at the start. The conversion rules prohibit the use of complex constructs and complex data types. However, for register accesses, this is a very minor constraint.

Although this is not the most elegant approach, it can be successfully implemented and has the advantage that the entire simulation environment is coded in Verilog, requiring no cosimulation tools. It is important to note, however, that the C to Verilog translation should be automatic. Manual translations are error prone and, consequently, can slow down the system verification phase by introducing unintended behavior.

Before an approach such as this is adopted, it requires thorough analysis by representatives of the software and hardware teams. The software team will require a well-defined programmer's reference guide for the ASIC earlier than would otherwise be the case. The programmer's guide defines all the register addresses and register field contents. The ASIC team may need to request additional software driver functions for register setup sequences, over and above those that would normally be used by the software team. For example, the ASIC may contain optional extra functionality for future use that is not being used in the initial version of the product and, as such, would not be exercised by the low-layer software driver. However, the ASIC team will want to have this extra functionality tested, so they will require the developers of the software driver code to include functions to activate these parts of the circuit. The benefit to the software team is that their low-level software layer is at least partly tested by the ASIC team and is tested well in advance of when it could otherwise be accurately tested.

5.4 Extending the Simulation Strategy

A number of EDA companies have extended the ideas presented in the simulation strategy in the previous section through the promotion of dedicated verification tools. These tools allow testbench drivers (sometimes called *transactors* or *BFMs*) to be controlled using higher-level commands or high-level languages. Some vendors have extended VHDL, Verilog or C++ to provide specific verification constructs, where others have proprietary hardware verification languages (HVLs). Some tools have built-in code coverage analyzers to determine how much of the design has been simulated. Some can even automatically create test cases for code that has not been exercised with existing tests. There is also a growing number of tools that will analyze the code for errors without the need to run simulations. These include tools such as formal verification that convert register-transfer-level (RTL)/gate-level code into mathematical equations allowing comparison between different netlists. Tools are discussed in Chapter 14, "Design Tools," but it is recommended that the reader visit EDA company Web sites to get the latest information.

These tools are particularly useful for large and system-on-a-chip (SoC) designs, where the overhead in cost and initial setup and training will be quickly recovered by the earlier availability of a high-quality test environment. Before deciding on a new verification tool, the project team must ensure that the tool is suitable for testing any third-party IP blocks and that it supports cosimulation with any embedded software. Training for the team is essential to ensure the smooth introduction of any new tools into the design flow.

5.5 Reducing Top-Level Simulation Run Times

As the size of ASICs continues to grow, reducing simulation times is increasingly important. A number of techniques are explained in the following sections.

5.5.1 Increasing Workstation Performance

The simplest way to increase simulation speeds is to increase the simulation host workstation performance. Replacing workstations with newer ones that have faster, newer processors or increasing the amount of available local memory (SRAM) can have a significant impact on simulation times. On most machines, the swap rate and CPU usage can be analyzed, and this will show whether the memory or CPU need upgrading.

The increase in simulation speeds is most needed during RTL chip-level tests and gate-level tests. Machines and memory can be rented for short periods if the project budget does not justify the outright purchase of the equipment. If renting workstations or additional SRAM is a realistic possibility, ensure that it is ordered well in advance. These are specialist items and often cannot be easily sourced at short notice.

5.5.2 Changing Simulation Tools

Different types of simulators will run the code at varying speeds. The traditional starting point for ASIC simulations is the interpreted event-driven simulator. This combines relatively fast compile times with acceptable run times until the ASICs reach a certain size threshold. However, alternative simulator tools, such as cycle-based simulators or hardware accelerators, are available, and although they have longer compile times than do event-driven simulators, these can significantly reduce test run times for large designs. These tools are discussed in more detail in Chapter 14, "Design Tools." The new style simulation tools typically perform best when both the testbench and the ASIC are synthesizable.

5.5.3 Analysis of Simulation Statement Executions

The simulator host machine takes a finite amount of time to execute the code statements in each module. Code coverage tools can identify those modules whose statements or functions are most frequently executed. These could be either statements or functions that take little time to execute once but are exercised frequently throughout the test or those that take a long time to execute once, for example, those that do large text I/O transfers. The testbench engineer can analyze the information provided by the code coverage tools and may find a way of restructuring the tests or recoding certain modules so that they execute faster.

For modules with large text output, the transfers can be reduced by outputting the data only when trying to identify bugs.

5.5.4 Preloading RAMs

ASICs often have internal or external RAMS that store information used when processing data. The RAMs are typically accessed via the CPU model, but these accesses can be slow in the simulator. Each write, for example, may take three or four system clock cycles. For tests requiring preloading of large sections of RAM with predefined data, this type of CPU-driven loading is unnecessarily slow. Many simulators allow the option of direct access to the RAMs, allowing loading of all bytes of information "instantaneously." Dumping the contents of the RAM instantaneously can also be useful when debugging problems.

5.5.5 Using Behavioral Models/Test Modes

Sometimes, a small number of RTL-coded modules are responsible for slowing down simulations significantly. For example, a digital-phase locked loop that runs at significantly higher clock speeds than the rest of the chip, and consequently generates thousands of extra simulation events, will inevitably slow down the simulation considerably. When running top-level simulations, this type of module can be replaced with behavioral models that run significantly faster. Alternatively, test modes can be used to bypass some parts of the chip, which are replaced by empty dummy modules.

5.5.6 Running Tests in Batch Mode

Developing scripts and makefiles that enable the tests to be run in batch mode takes a significant amount of effort to develop at the start of the top-level simulation phase. However, this effort will pay dividends throughout the project because the tests are often rerun many times as bugs are identified and fixed. Makefiles are particularly useful because they will recompile only out-of-date files.

5.6 Speeding Up Debugging

Simulators have a range of features that facilitate identifying and resolving bugs in the design. Examples of these features include the ability to save simulation snapshots and source code debugging.

5.6.1 Saving Simulation Snapshots

Most simulators have the ability to freeze the state of the simulation during the test. When the code contains a problem that manifests itself only after a significant simulation period has elapsed, it can take a considerable amount of time to identify the bug if multiple simulation runs are needed. For example, on each successive run, several more waveforms may be traced to narrow the problem down.

In these cases, the simulator should be rerun until the simulation time approaches the region where the fault condition arises. If the state of the simulation is saved, the test can be restarted from this saved state or snapshot if the bug is not identified immediately.

Once the bug is found, the potential "fix" sometimes can be tested quickly by restarting from the saved state, then using the simulator's *force* command to override signal values. If the *force* solution seems to work, the code is then modified with a real fix, and a full simulation is reexecuted to confirm that the problem has been properly resolved.

5.6.2 Source Code Debugging

Many simulators for both VHDL and Verilog include source code debuggers.

Being familiar with these tools will improve the time to resolve problems in the design. These debuggers allow the setting up of break points and enable single-stepping through each line of code. Some features are particularly useful, such as breaking every time a process is run and having the ability to step backward in time following a break point. Each member of the team should receive a sufficient level of training on the debugger to be familiar with its most useful features.

5.7 Different Types of Testing

It is very important for all project engineers to understand the goal of the testing done at the different levels. This ensures that each module or submodule is tested in an appropriate way at each level, which, in turn, should reduce the number of bugs that slip through to the top-level simulations.

5.7.1 Module Testing

The goal of the module testing is to check exhaustively that each function, configuration and option of the module operates as specified. The tests should aim for 100% statement, branch and toggle coverage. These tests should be performed on both RTL and gate-level descriptions of the module. Testing exhaustively at this level has a number of advantages:

- Because only one module is being simulated, the tests will run faster, and bugs can be identified quicker.
- Testing thoroughly at the module level derisks the subsystem and chip-level simulations.
- It may not be possible to exercise all code branches for the module later in the top-level simulations when the module inputs are less controllable.

Individual module testing is normally based on custom testbenches developed by the designer of the module in question. Adapting system testbench drivers to generate input stimuli may be an option, but this is not likely to be the case. The system testbench components will probably not yet be in place. Additionally, if the module under test is deeply embedded in the ASIC, well away from the chip I/O boundary, a top-level driver could require complex translation layers to convert its interface to drive the interface of the module under test. It is, therefore, likely that low-level custom drivers and monitors unique to the module under test will have to be developed.

The first stage in module testing is to develop a formal module test specification that lists all the tests. The objective of each test should be explained, and the measure by which it will be deemed to have passed or failed should be defined. Ideally, the tests should be self-checking to allow quick rerunning of the tests at the module level if any subsequent code changes are required in the module.

After a number of RTL code simulations have completed, gate-level simulations should be run. This ensures that the code has been written correctly for synthesis and that the synthesis tool has generated a netlist that matches the RTL code. The results of all the simulations, including the percentage test coverage achieved in the module, should all be made available as input to design reviews.

For complex modules with complex interfaces, generating an exhaustive testbench at the individual module level can take as much effort as designing the module to which the UUT is interfaced. In these cases, it is probably worth doing only limited testing at the individual module level and, instead, concentrate on testing the module thoroughly in a subsystem environment. The subsystem environment groups the module with several other closely related neighboring modules into a subsystem and tests all these modules together as one larger unit with drivers and monitors tailored to the subsystem environment.

5.7.2 Subsystem Testing

The goal of subsystem simulations is to run tests on a number of modules that combine together logically to form a single higher-level function and that are more efficiently tested as a combined unit, rather than as individual subunits. There is a number of advantages to running subsystem simulations, rather than just chip-level simulations:

- The subsystem simulations can typically start earlier in the project. This will reduce design risks because any problems will be found earlier, allowing more time to implement changes without impacting the sign-off date.
- Frequently, all the chip-level simulations rely on certain subsystems functioning properly before any other feature can be tested. Subsystem simulations can verify any such functions before the start of chip-level simulations.
- The subsystem simulations include less code and will simulate faster.
- Some modules are difficult to test at the chip level when the inputs to the module are difficult to control. For example, consider a UUT whose function is to analyze the status of data packets received in a datacomms chip. The input to the UUT in the chip-level environment can change only as fast as the minimum size packet is transferred. In the subsystem simulation, the size of packet is irrelevant. The interface can be modeled just as accurately

in a few clock cycles as it can in the thousands of cycles it would take to model a full packet transfer. This significantly reduces the amount of clock cycles needed to test the UUT fully.

The testbench for subsystem simulations should be based as closely as possible on the top-level testbench. Ideally, the subsystem testing can use similar configuration files, input files, testbench drivers and testbench monitors as those used in the chip-level testing. This saves time during chip-level testing because the input files used at the subsystem level may need only minor changes to be reused at chip level.

5.7.3 Chip-Level Testing

The goal of the chip-level simulations is to prove that the fully integrated netlist performs the functions or algorithms defined in the original specifications. The simulations should exercise the interfaces between the modules and should test all features that could not be tested in isolation at the module or subsystem levels.

The tests that should be carried out at the chip level can be classified into a number of distinct categories:

- Performance testing checks that the ASIC satisfies its performance requirements and can process data at the required throughput rates.
- Compliance testing tests that the ASIC conforms to any protocol or specification that it was designed to conform to.
- Boundary or corner testing tests the ASIC's response to unusual or unexpected input conditions. Some such input conditions may not themselves be legal. However it may, nonetheless, be an operating requirement that the chip responds in a noncatastrophic way to such input combinations.
- Stress or load testing checks the ASIC's ability to perform as expected under maximum worst-case load conditions. For example, worst-case load conditions in a 12-port full duplex Ethernet switching ASIC probably means that all 12 ports are simultaneously transmitting and receiving minimum-sized data packets with minimum interpacket gaps, with all statistical analysis modules switched on and enabled.

5.7.4 Gate-Level Testing

After RTL testing has been carried out at the chip level or after at least some RTL testing has been completed, gate-level simulations can begin. Gate-level tests are typically comprised of a subset of the chip-level tests described in the previous section.

They are intended to prove that the design was synthesized correctly and that the required timing cycle constraints have been met. Initial gate-level simulations can use unit delay modeling, whereas subsequent gate-level simulations should use estimated cell delays, based on timing library models. The gate-level simulations using timing library delays help to prove that the ASIC will function correctly after fabrication when physical cell and interconnect delays come into play. They should be run with best-case and worst-case timings. Static timing analysis should be used to check that all timing requirements have been met.

5.7.5 Postlayout Testing

The aim of post layout testing is to prove that the netlist still functions correctly when the real signal delays, as determined by the layout, are back-annotated into the simulation environment. Back-annotation is done using the standard delay format (SDF) files generated in the layout process. This helps to prove that the chip will work when it is fabricated using the current layout. ASIC sign-off should be delayed until postlayout simulations have been completed. Particular attention should be paid to setup and hold violations during these simulations. The postlayout simulations should be rerun with back-annotated delays for minimum, typical and worst-case timings. These simulations take a long time to complete so, typically, only a subset of the gate-level tests are rerun. As with prelayout checking, static timing analysis should also be used at this stage to verify timings.

5.7.6 Board-Level Testing

The goal of board-level testing is to prove that the ASIC will function correctly with the other components in the target printed circuit board (PCB) environment. Only a subset of the chip-level simulations need to be tested at the board level. These tests also prove that the PCB has been designed correctly and that the interfaces between different chips were properly specified and correctly interpreted.

5.8 Generation of ASIC Test Vectors

There is a number of different types of test vectors that need to be generated and tested.

Functional test vectors can be generated and captured automatically in the chip-level simulation environment. This assumes that the chip-level drivers/monitors have been written to conform to the requirements of the ASIC vendor's tester tools. These requirements include inputs being switched in certain timing windows in the clock cycle, minimum required times to change the direction of buses, defined times for strobing the outputs of the ASIC, etc. The ASIC vendor should be consulted early in the project to establish the exact requirements.

Parametric tests are normally a small set of vectors that drive the ASIC and the ASIC I/O into a number of defined states: high, low and tri-state. These are best done automatically using the testbench.

The timing tests prove that the ASIC can run at the required application speed. Most functional test vectors are not run at the speed at which the device will operate, due to speed limitations imposed by connecting to the testers.

Each of the above sets of test vectors must be simulated for best-case, typical and worst-case timings. The chip-level test environment should be able to run these vectors. The test vectors must be rerun with back-annotated delays. Due to the number of vectors that must be applied to the chip, test vectors can take a very long time to run. It is usually possible to run the simulations across multiple machines to speed things up. Alternatively, many vendors will do the simulations themselves but at additional cost.

5.9 Summary

As ASIC sizes continue to grow, the design of the top-level testbench is often one of the most crucial tasks in the project. It should be specified and architected at the start of the project so that it can be designed and debugged in time for the subsystem and system simulation phases. The testbench design team should observe a similar design flow as that used in the design of the ASIC itself. The wider team should assist in the process of identifying the list of required testbench features and consider how all the necessary input stimuli can best be generated.

The principal components in a testbench are the input drivers, the UUT and the output monitors. A master configuration file can control the main characteristics of a given test. Fields in the configuration file determine which drivers and monitors are enabled. The operating mode of the ASIC is usually determined by the commands that are set up for the CPU driver. These commands are converted into ASIC register read/write transactions.

A good testbench driver will allow input stimuli to be generated through a series of relatively high-level commands supporting a number of parameters. This type of interface enables setting up complex input patterns in a very short timeframe. Drivers that model data sources sometimes allow the user control over the selection of the data patterns used.

There is a number of steps that can be taken to speed up aspects of the simulation process. Makefiles minimize compilation times by recompiling only the source code

modules that have changed. Judicious use of configuration files and input data files for
the drivers means that multiple different types of tests can be run without the need to
recompile the testbench—the input file changes determine the nature of the test. Time
invested in making testbenches self-checking is usually well worth while. Self-checking testbenches, if properly designed, are inherently more reliable than those requiring
manual checking, and they speed up the process of regression testing considerably.
Simulations can also be speeded up by increasing the processing power of the simulation platform or, in some cases, by using alternative simulator technologies, such as
cycle-based simulation or hardware acceleration.

   There are several different levels of simulation, from individual module testing
through chip-level testing to board-level testing. The ASIC team should understand the
scope and objective of each of these different levels to ensure that each level is
approached in an optimal way to produce the required results.

Synthesis

6.1 Introduction

Synthesis is the process of translating a register-transfer-level (RTL) description into a gate-level netlist using a synthesis tool. Interest in synthesis tools surged among the wider design community in the late 1980s and early 1990s as design sizes started to grow beyond the levels that were practically achievable by traditional manual translation methods. Synthesis tools were seen as a way of creating larger designs with acceptable development times.

Within the ASIC domain, the Synopsys synthesis tools have traditionally dominated the industry. Its tools are still delivering high-quality results, supporting a wide range of language constructs and a rich suite of user interface commands. However, at the time of writing, Cadence, another large EDA tools provider, had begun to establish itself as a significant contender in the synthesis tools market following its purchase of Ambit, an independent synthesis tool developer. Several other vendors, such as Compass Design Automation, Mentor Graphics, and Avanti, also provide synthesis tools, but we focus on Synopsys and Cadence offerings in this chapter to limit the scope of the coverage.

This chapter serves as an introduction to synthesis terms and concepts. The complexity of synthesis is such that an in-depth coverage is outside the scope of this book. However, the intention is that this chapter should leave the reader with a reasonable understanding of the most important synthesis issues. It combines relevant background material with sample script excerpts for two popular synthesis tools.

6.2 The General Principle

Synthesis tools begin by converting an RTL description into a generic technology-independent format. In this phase, there is some basic optimization carried out. The generic database is then mapped to a technology-based gate-level representation, using a vendor's technology library. Subsequently, the gate-level representation is optimized to minimize area, speed and power, according to user-specified design constraints. Traditionally, following layout, physical constraints were fed back into the synthesis tools for fine-tuning and reoptimization. However, with the latest range of synthesis tools, layout-dependent physical constraints can be incorporated into the synthesis process from the outset. This reduces the number of iterations required to achieve timing closure.

6.3 Top-Down versus Bottom-Up Synthesis

Early versions of synthesis tools were limited by the size of designs they could reasonably tackle in one synthesis run. In the early 1990s, a figure of 3,000–5,000 gates was recommended as a sensible upper limit on the maximum size that should be handled by the synthesis tool. The practical bounds were, of course, influenced by the power and memory limitations of the processing platform. As design size increased beyond relatively small limits, the processing time started to increment exponentially.

For this reason, larger designs were synthesized in a bottom-up approach. Designs were broken up into synthesizable chunks, which were synthesized independently. The individual designers would agree on mutually acceptable time budgets on the interface signals between their blocks, typically allowing a small timing margin to allow for the insertion of driving buffers or larger-than-predicted loads when the blocks were connected. It was, and still is, common in this type of approach to mandate that all or nearly all outputs from each module are registered to allow the maximum possible timing margin on such output signals entering other blocks as inputs.

This bottom-up approach has a number of disadvantages. It takes time to agree on timing budgets between individual blocks. If the timing margin allowed between blocks is insufficient, the top-level timing requirements will not be met. If the margins are excessive, the gate-count and power consumption will be higher than necessary. After individual synthesis runs are completed, there is often a requirement for iterative synthesis runs where a higher-level "characterization" of the input and output timing and driving characteristics are used to set the input and output constraints for resynthesis instead of the manually agreed-on starting budgets. This approach is time-consuming, and the synthesis scripts can become quite cumbersome. The bottom-up approach normally means that all designers must have some proficiency in synthesis, as opposed to one synthesis expert synthesizing all blocks in one go in the top-down approach.

With the advent of the SoC era and designs with ever-increasing gate counts, the demand for higher-performance synthesis tools increased. In the late 1990s, prior to its purchase by Cadence, Ambit announced its BuildGates product, which was capable of synthesizing more than 100,000 gates in a top-down style. This was a significant new development and allowed full chip synthesis in one top-down run in certain cases. The competitors, including Synopsys, subsequently announced and delivered tools with similar capabilities. However, the size of many of today's designs still exceeds the maximum practical top-down capacity of any of the synthesis tools currently on offer. Additionally, a new problem associated with higher gate counts has driven the current direction of synthesis tool developments.

6.4 Physical Synthesis Tools

At the time of writing, many designs under development exceed 1,000,000 gates. These design sizes have been made possible by significant developments in silicon technology, where physical gate geometries are continually shrinking. As gate geometries shrink further and gate speeds consequently increase, the delays in a circuit are no longer dominated by the intrinsic delay through the gates themselves. Gate interconnect becomes a significant element of the overall delay equation. The problem with the traditional synthesis tools and technology library models is that, so long as they do not take physical layout and placement into account, they cannot accurately model the interconnect delay, which is heavily dependent on physical layout. Traditional synthesis relies on relatively crude wire-load modeling to model the effect of cell placement and interconnect routing in advance of knowing what the placement is going to be. Wire-load models are used to approximate the load on a driving cell by looking at the number of fanouts and taking a statistical estimate of the wire length, based on the size of the block for which the model applies. Technology libraries contain tables of wire-load models from which loading estimates are derived during synthesis, based on the size of the block being synthesized. These estimates are used for prelayout synthesis. After layout has taken place, the physical layout information can be back-annotated into the synthesis tool, and this information can be used to verify that all timing constraints are still being met or, if they are not (as is normally the case), it can be used as an input to a resynthesis process. This is an iterative and, therefore, time-consuming approach, and the degree to which it is successful is sometimes difficult to predict. Furthermore, as geometries continue to shrink, this approach becomes less and less practical as the mismatch between prelayout and postlayout timings diverge further.

At the time of writing this edition, yet another new generation of synthesis tools is hitting the market. These attempt to address the limitations of the traditional synthesis approach by taking physical layout into account during logic synthesis. For the purposes of this book, they are referred to generically as *physical synthesis tools*. The Synopsys offering, Physical Compiler, is built on its Design Compiler product, whereas the Cadence offering, Envisia PKS (Physically Knowledgeable Synthesis), is a superset of the Ambit BuildGates logic synthesis tool. The general principle of physical synthesis tools is that they treat layout as an integral part of the synthesis process, instead of the traditional view of considering layout as a separate postsynthesis step. This cuts out or greatly reduces the iterative and time-consuming manual process of back-annotating information from a separate layout process into a separate logic synthesis process. Synopsys refers to the concept of "unifying synthesis and placement" in its Physical Compiler tool. Cadence's Envisia PKS leverages the close correlation with its Silicon Ensemble family of placement and route tools. A further advantage of these tools is that they not only help achieve accurate timing closure much more quickly but can also result in significantly better overall results in terms of timing margins, gate counts and power consumption figures. By taking layout information into account at the outset, they can influence the architecture to achieve the best results. In the traditional synthesis approach, only limited rearchitecture can take place when the starting point of resynthesis is a gate-level implementation. When the starting point is RTL, much more fundamental and, consequently, more effective rearchitecture is enabled.

Physical synthesis tools are gaining in popularity and are a necessary improvement for processing deep submicron designs.

6.5 Scripts versus GUIs

The main synthesis tools can be driven via a GUI or, alternatively, via command script files. As a general rule, beginners will use the GUI but quickly move to using scripts. The GUI, being menu-driven, is easier to use at first. The GUI is sometimes also valuable to experienced users when tracing very specific paths in a netlist or examining the chosen architecture that the synthesis tool has used to implement a given piece of logic. GUIs also have useful features, such as the ability to highlight critical paths in the circuit and to link these back to the source code. The critical timing path in a circuit is the logic path with the largest timing violation of its allowed timing budget or, if there are no violating paths, it is then the path with the least timing slack, when compared with its allowed timing budget.

Once users become familiar with scripting, they tend to revert only occasionally to using the GUI and then only for very specific purposes. Scripting allows more powerful usage of the tool's command set than the GUI permits (e.g., command looping). Additionally, the scripts can be executed overnight and on weekends, unlike driving a GUI, which requires user interaction. Where synthesis tool providers provide separate licenses for GUIs and script-driven compilers, it is advisable to purchase a minimum number of GUI licenses and a larger number of non-GUI licenses.

6.6 Common Steps in Synthesis Scripts

The main synthesis tools are command and feature rich. There may be more than 100 commands and multiple options and parameter settings for many of these commands. Many designs require only a small subset of the total available command set. The following section looks at some of the more common commands and their sequence in a typical synthesis script. First, let's examine a typical sequence of actions in a synthesis script. These are described in the form of generic action statements with some sample Synopsys Design Compiler and/or Ambit BuildGates syntax formats given as examples. Later, we will look at stand-alone Design Compiler and BuildGates standard script examples. Note that the sequence may vary from tool to tool and according to user style. Additionally, this is a relatively simplified view of the sequence that leaves out many of the complex actions that may be required in more complex designs. The purpose here is simply to provide an introductory overview.

6.6.1 Sample Script Action Sequence

Set global variables. This covers general settings that affect the running of the tool. For example, in BuildGates, the command *set_global echo_commands true* causes each command to be echoed on standard output prior to execution.

Set user-defined variables. Users can define their own variables to make the scripts more readable and more concise. For example, a long pathname to a project directory can be mapped to a user-defined variable, which is more convenient to refer to each time it is required subsequently in the scripts. The keyword *set* is required by BuildGates, for example:

```
set  design_root    /myteam/comms_products/projects/
     modem_project
set  scripts_area   ${design_root}/scripts
```

Set environmental and operating conditions. The synthesis tool needs to know the operating conditions, because these are effectively design constraints. For example, circuit delays are affected by the operating conditions, such as voltage, temperature and process parameters. In the following Design Compiler example, some operating conditions are defined. The first three lines create user-defined variables (OPER_COND, DES_LIB and WIR_MODEL), which are set to predefined values supported in the technology library. The two lines following these apply these values to help define the synthesis operating environment.

```
OPER_COND = WCMIL
DES_LIB   = dl2000
WIR_MODEL  = wm300e
set_wire_load  WIR_MODEL -library  DES_LIB
set_operating_conditions  -library  DES_LIB  OPER_COND
```

Wire-load models are a method of estimating the impact of interconnect length and fanout on circuit path delays. The reader is referred to the synthesis tool user manuals for a fuller explanation. It is, however, interesting to note that, in deep submicron designs, it is largely because of the imprecision of wire-load modules at these geometries that it has been necessary to rely on real physical layout implementation in the new-generation physical compiler tools.

Read in required libraries. The libraries include the basic target technology cell library (AND gates, OR gates, F/Fs, LATCHES, etc.), and, where appropriate, any additional libraries, such as I/O cell libraries and memory cell libraries. In BuildGates, the following command shows one of a number of commands for reading in a technology library.

```
read_ctlf   -process 0.5 -voltage 5.5 -temperature 85
    ${libraries_area}/dl2000.ctlf
```

Read in required design files. This step reads in the source code that is about to be synthesized. If this is a first-pass synthesis script, the source files are normally in Verilog or VHDL format. If it is an intermediate synthesis script, the source code may already have been translated into the synthesis tool's native database format (e.g., *.db* for Design Compiler or *.adb* for BuildGates).

For original source code in Design Compiler, we could have:

```
read -format verilog {  BitSampler.v  BitSlicer.v  }
```

or in BuildGates:

```
read_vhdl  BitShifter.vhdl
```

Define clocks. Once a design contains clocked logic elements, the clocks must be defined. The clocks then become the timing reference for nearly all other signals in the design. For example, inputs are specified with respect to the clock timing of the clock domain(s) into which they arrive, and outputs are specified relative to the clock domain(s) in which they are generated. An exception to this is purely combinational

input to output delays. Design Compiler and BuildGates define clocks in subtly different ways. The example below shows a Design Compiler clock definition. The
dont_touch_network command instructs the compiler not to buffer this net, even if it
does not appear to have the required drive strength.

```
create_clock   -name HsClk   -period 100 -waveform {0 5}
dont_touch_network   find(clock,"HsClk");
```

Set input and output timing constraints. The examples below show a Design
Compiler syntax for defining some input and output delays. Note that the interpretation
of the *set_output_delay* command is a common source of error. Intuitively, one might
think that the *set_output_delay* command in the example below is setting a requirement
that the outputs are stable 33 nsec after the *HsClk* active edge. However, it is, in fact,
setting a requirement that the outputs are stable 33 nsec before the next active *HsClk*
edge or 67 nsec after the previous edge, in the case of a 100-nsec clock period.

```
set_input_delay -max 10 -clock "HsClk"   all_inputs( )
set_output_delay -max 33 -clock "HsClk"   all_outputs( )
```

Set multicycle paths and false paths. In many ASICs or ASIC modules, there
are a number of inputs that are either permanently tied to one logic level or that settle to
their stable value so far in advance of the cycle in which they are used that they are
never going to pose a timing problem. A common example of the latter is a CPU register-driven control signal. To prevent the synthesis tool from unnecessarily trying to
minimize path delays from these signals, they can be set as *false* timing paths. The
same is also true of timing paths to various outputs, which are known to settle to a stable value so far in advance of when they are used that the path delay is irrelevant.

A second saving of processing effort for the synthesis tool arises in what are
known as *multicycle* paths. These are signals that are set up one or more cycles in
advance of when they are needed but not so far in advance that they can be ignored
completely, as is the case with *false* paths.

Setting false and multicycle paths has a number of advantages; the gate count
should be reduced, which, in turn, reduces power consumption, and the layout tool can
concentrate on genuine worst-case timing paths when timing-driven layout is used.
However, specifying too many false and multicycle timing paths can slow down compile times.

The example below illustrates sample Design Compiler syntax for these commands.

```
set_multicycle_path 2 -setup -from muxSelect   -to dataOut
set_multicycle_path 1 -hold -from muxSelect   -to dataOut
set_false_path   -from   testEnable
```

Note that the absence of a *-to* clause in the *false path* example implies to *all* possible endpoints.

Set input driving constraints and output loading constraints. Path timing calculations are affected by the drive strength of the circuit inputs and the expected load on any output signals. In the Design Compiler examples below, the input driving strength for all inputs is set to that of a *D*-type flip-flop's *Q*-output. The output load on all outputs is set to that of the *A*-input of a standard inverter cell using the *load_of* function.

```
set_driving_cell  -cell DFF2  -library DES_LIB  -pin Q
    all_inputs()
set_load   load_of  (dl2000/IV/A)  all_outputs()
```

Set fanout and slew constraints. In the BuildGates example below, the *set_fanout_load* example enforces a design rule check on the fanout of the signal *alarmOn*. The *set_slew_time* command lets the tool know that the input signal *dataIn* does not have ideal rise and fall times. The tool will take the slew parameters into account when calculating path delays.

```
set_fanout_load 3  [find -port -output alarmOn]
set_slew_time 0.5 dataIn
```

Run precompile design checks and generate precompile reports. The *check_timing* command in BuildGates runs a number of consistency checks on timing constraints in the design. It can point out missing or inconsistent constraints before investing significant time in a wasted optimization run.

Compile the design. This step is sometimes referred to as *optimizing the design* and sometimes referred to as *compiling*. In Design Compiler, there is a single compile command with many possible parameters. In BuildGates, the user is allowed access to a variety of so-called transforms, allowing more flexibility but probably requiring more user understanding. The example shows a Design Compiler syntax with a sample selection of parameter settings.

```
compile  -boundary_optimization -map_effort  medium  -scan
```

Generate output reports. After a compilation or optimization run, it can be useful to request that status information be reported by category into user-named output files. Some Design Compiler reporting commands are shown below. One of the most important reports is the *violation* report, which gives a list of all violated user-constraints.

```
report_design >  design.rep
report_area >   area.rep
```

```
report_timing > timing.rep
report_constraints -all_violators > viol.rep
```

Generate the output netlist in required formats. When a synthesis run is complete, the resultant netlist can be saved in a variety of formats. For example, the BuildGates tool can save the netlist as a Verilog, VHDL or *.adb* file. The example shows the Verilog hierarchical option.

```
write_verilog  -hierarchical  ${design_root}/verilog_gates/
    mydesign.v
```

6.6.2 Sample Scripts

Having looked at the general sequence of steps in a synthesis script, we now look at sample scripts for the Synopsys Design Compiler and Cadence-Ambit BuildGates synthesis tools. Again, the sample scripts are relatively simple and do not use many of the more powerful features available in both of these tools. The reader is referred to the user guides and application notes of the specific synthesis tools for more in-depth coverage. Note that the scripts are not intended to be equivalent. In fact, they are deliberately different to show a wider range of features. The Design Compiler example assumes a design with Verilog source code, whereas the Ambit example assumes a design based on VHDL source code. Finally, neither script is fully complete. They are intended only to show a sample of the typical features and possible action sequences that might be encountered in a true application script.

6.6.2.1 Design Compiler Sample Script

```
/* Note: target libraries and link libraries are set-up in
    the  */
/* .synopsys_dc.set-up file, which is read in automatically
    */

/* Vendor set-up */
/* Source the vendor's set-up file which sets various flags
    */
/* and global variables to vendor-recommended values */
include vendor_dl2000_set-up.dc

/* Set some operating conditions */
OPER_COND = WCMIL
DES_LIB   = dl2000
WIR_MODEL  = wm300e
set_wire_load WIR_MODEL -library DES_LIB
set_operating_conditions -library DES_LIB OPER_COND
```

```
/* Exclude the use of a particular flip-flop cell */
set_dont_use (dl2000/FF5)

/* Set user-defined variables */
MCLK_PERIOD = 20
GENERIC_INPUT_DELAY = 8
GENERIC_OUTPUT_DELAY = 14

/* Read in the design */
/* (mix of Verilog and pre-synthesized .db modules) */
read -format verilog ../rtl/frontEnd.v
read -format verilog ../rtl/dataReg.v
read -format verilog ../rtl/cpuif.v
read -format verilog ../rtl/modem.v
read -format db   ../dbfiles/bitSlice.db

/* Select a design to work on */
current_design = modem
check_design > modem_check.rep

/* Link the design */
link

/* Ensure that the pre-synthesized bitSlice block is not
    re-optimized */
dont_touch find(cell,bitSlice)

/* Set up the scan test configuration */
set_scan_style  lssd
set_test_methodology  full_scan

/* Define clocks */
create_clock -period MCLK_PERIOD -waveform {0 MCLK_PERIOD /
    2} find(port,"MClk")
dont_touch_network find(clock,"MClk")
set_clock _uncertainty  -set up  0.25 find(clock,"MClk")
set_clock _uncertainty  -hold  0.25 find(clock,"MClk")

/* Apply timing constraints */
set_input_delay GENERIC_INPUT_DELAY -max -clock "MClk"
    all_inputs()
set_output_delay GENERIC_OUTPUT_DELAY -max -clock "MClk"
    all_outputs();
set_false_path -from testEnable -to ramData*
```

```
set_multicycle_path 2 -set up -from frontEnd/cntr_reg/CP -
    to ramDataIn
set_multicycle_path 1 -hold -from frontEnd/cntr_reg/CP -to
    ramDataIn

/* Apply cell load and driving constraints */
set_load   2 * load_of(dl2000/IV/A) all_outputs()
set_driving_cell  -cell FF2 -pin Q -library dl2000
    all_inputs()

/* Compile the design   */
compile  -map_effort  medium  -scan

/* Characterize each block and save characterized
    constraints */
remove_attribute find(cell,bitSlice) dont_touch
characterize bitSlice
current_design "bitSlice"
write_script > ../constraints/bitSlice.char
current_design "modem"

characterize frontEnd
current_design "frontEnd"
write_script > ../constraints/frontEnd.char
current_design "modem"

characterize cpuif
current_design "cpuif"
write_script > ../constraints/cpuif.char
current_design "modem"

characterize dataReg
current_design "dataReg"
write_script > ../constraints/dataReg.char
current_design "modem"

/* Compile each individual subblock with characterization
    data */
current_design dataReg
set_scan_style lssd
set_test_methodology full_scan
include ../constraints/dataReg.char
ungroup -all
compile -incremental_mapping -map_effort high > ../reports/
    dataReg.rep
report_timing >>  ../reports/dataReg.rep
```

```
report_area >>  ../reports/dataReg.rep
report_constraints -all_violators >>  ../reports/
    dataReg.rep
write -format db        -output ../dbfiles/dataReg.db
write -format verilog -output ../verilog_files/dataReg.v
current_design modem
...
... repeat same for other modules ...
...

/* Finish with incremental top-level compile */
set_scan_style lssd
set_test_methodology full_scan
compile -incremental_mapping

/* Save output files */
write -format db -hierarchy -output ../dbfiles/modem.db
write -format verilog -hierarchy -output ../verilog_out/
    modem.v

/* Generate reports */
current_design = modem
report_design >  ../reports/modem_design.rep
report_area >   ../reports/modem_area.rep
report_timing > ../reports/modem_timing.rep
report_constraints -all_violators > ../reports/
    modem_viol.rep
report_constraints -all_violators -verbose > ../reports/
    modem_viol_verb.rep

quit
/* END SCRIPT */
```

6.6.2.2 BuildGates Sample Script

```
#  Set user-defined variables
set project_dir  /myproject/mac_core
set synth_dir ${project_dir}/synth
set vhdl_work_area ${synth_dir}/vhdl_work

#  Read in target library
read_alf ${synth_dir}/target_lib/dl2000.alf
set_global target_technology dl2000

#  Set operating conditions
set_operating_conditions WCCOM
```

```
# Set 'dont_utilize' on some library cells to prevent their
    being used
set_cell_property dont_utilize true -lib dl2000 [ get_names
    [find -cellref ff3s*] ]

# Compile standard VHDL packages
set_vhdl_library IEEE        ${synth_dir}/ieee
set_vhdl_library DL2000      $corelib_lib_area
set_vhdl_library MAC_LIB     $vhdl_work_area
set_vhdl_library WORK     MAC_LIB
set_global hdl_vhdl_environment standard

# Read in VHDL design source files
read_vhdl ../sources/myDefsPkg.vhdl
read_vhdl ../sources/crcGen.vhdl
read_vhdl ../sources/statistics.vhdl
read_vhdl ../sources/rx.vhdl
read_vhdl ../sources/tx.vhdl
read_vhdl ../sources/speedSense.vhdl
read_vhdl ../sources/clocks.vhdl
read_vhdl ../sources/EtherMac.vhdl

# Define Clocks
set_clock_propagation ideal
set_clock clk_ideal  -period 9 -waveform {0 4.5}
set_clock_uncertainty  -late 0.3
set_clock_uncertainty  -early 0.3
set rxtx_clock_pins {
                EtherMac/clocks/txClk
                EtherMac/clocks/rxClk
                }
set_clock_arrival_time -clock clk_ideal -rise 0 -fall 4.5
    [find -pins $rxtx_clock_pins]

#  Apply Constraints
set_data_arrival_time -clock clk_ideal 3 [get_names [find -
    ports -input -no_clocks *] ]
set_external_delay -clock clk_ideal 3 [find -ports -output
    *]

#  Multi-Cycle Paths
set_cycle_addition -to EtherMac/rx/statistics/mux_buf*/D -
    late 1
set_cycle_addition -to EtherMac/rx/statistics/reg_load*/D -
 ·  late 1
```

```
set_cycle_addition -from testEnable  1

#  Execute first Ambit compile step
do_build_generic -design EtherMac

#  Generate a check report and save preliminary database
#  Subsequent runs can take this database as starting point
#  if no source code has changed
check_netlist -verbose >  ${synth_dir}/reports/
    EtherMac_generic.rpt
write_adb -hierarchical ${synth_dir}/adb/
    EtherMac_generic.adb
write_verilog -hierarchical  ${synth_dir}/vlog_out/
    EtherMac_generic.v

#  Execute 2nd Ambit compile step
#  i.e., map the design to target technology
set_current_module EtherMac
set_top_timing_module  EtherMac
source    ${constraints_area}/${top_module}_constraints.tcl
check_timing > ${synth_dir}/reports/EtherMac_timing.rpt
do_optimize -effort medium -no_design_rule
#  Save the database
write_adb -hierarchical ${synth_dir}/adb/
    EtherMac_mapped.adb
write_verilog -hierarchical ${synth_dir}/vlog_out/
    EtherMac_mapped.v

#  Execute optimization step
set_current_module EtherMac
set_top_timing_module EtherMac
do_xform_optimize_slack -effort high -incremental -
    critical_ratio 0.1
set_global  auto_slew_incr  true

#  Save the database
write_adb -hierarchical ${synth_dir}/adb/
    EtherMac_optimized.adb
write_verilog -hierarchical ${synth_dir}/vlog_out/
    EtherMac_optimized.v

# Generate timing report
report_timing > ${synth_dir}/reports/
    EtherMac_optimized_timing.rpt
```

6.7 Directory Structures

As is the case with the complete design database, it is useful to divide the synthesis directory into a number of different subdirectories. A carefully thought-out directory system makes it easier to find specific files as they are required. A consistent, company-wide approach to synthesis directories in terms of both types of subdirectories and their naming convention means that it will be easier to reuse the design and that it will be easier for a new project member to know where to find specific synthesis files when they need them.

There is no one correct best directory structure that can be recommended. However, it may be useful to use some of the following suggested subdirectory types:

- Sources: the source modules of the design
- Libraries: possibly a number of library files (e.g., the technology library files, the IEEE VHDL libraries, etc.)
- Constraints: synthesis constraint files
- Scripts: synthesis script/command files
- Reports: output reports (e.g., timing reports, design check reports etc.)
- Outputs: possibly a variety of output types, each meriting its own subdirectory, e.g., Verilog, VHDL, generic, etc.
- Back-Annotation: contains back-annotated timing information, e.g., SDF and set-load files

Note that it may make sense to have links to the real files in some of the subdirectories, instead of copying the files themselves over from a central server location. This may assist in maintaining stricter version controls or in minimizing storage requirements. However, if links are used, it is important to remember to include any relevant linked files if the synthesis database is being handed over to a third party outside the design team, e.g., the vendor, an IP customer, etc.

6.8 Special Cells

6.8.1 Handling Memory Cells

Memory cells can be behaviorally inferred in the source code, typically as two-dimensional arrays. Depending on the precise coding, these may translate into sets of flip-flops or latches. However, for anything other than very small memory requirements, this is an inefficient form of memory. Most vendors will supply precompiled memory cells with a wide variety of characteristics and sizes, which can be directly instantiated in the source code. These memory cells are significantly more compact than the "homemade" flip-flop types and should be used in preference, unless there is a good reason not to. From the synthesis point of view, once the memory cell is instanti-ated and a corresponding model is available in the library, there is no processing or decision-making required. The models must, of course, come complete with timing characteristics to enable timing paths to and from the memory cells to be analyzed. It is important to establish in advance with a vendor that synthesis-timing models will be available in sufficient time, in the case of custom memory cells.

6.8.2 I/O Cells

I/O cells can be inferred from an I/O library but this is an unusual approach. Fre-quently, the I/O cells are hand-picked by the designers from a set of options to meet very specific required characteristics. Characteristics under consideration might include slew rate, current drive, pull-up/pull-down termination options, voltage swing, Schmidt triggering, etc. For this reason, I/O cells are often instantiated directly in the source code.

6.8.3 Other Special Cells

Some designs will use precompiled macros provided by the ASIC vendor to speed up the design process. Many vendors provide their own or third-party precompiled macros or cores as an additional service in an effort to win the customer's business. There is a large and growing range of such macros available from most vendors. SRAMs, DRAMs, CAMs, MACs, and processor cores are but a few examples. These are generally all treated in the same way. They are instantiated directly in the code, and once a corresponding model exists in the active synthesis libraries, they are connected into the gate-level database during synthesis.

6.9 Miscellaneous Synthesis Terms, Concepts and Issues

Although the basic concept of a synthesis tool is simple, the implementations and realizations are very complex. There are usually numerous instructions available, with numerous options for many of these instructions. Additionally, there may be a large number of global variables used to guide the tool and get it to operate in specific ways. This text can give only a brief general introduction to synthesis and does not attempt to cover the full scope of this subject. However, in this section, a selection of some of the more common terms and concepts in synthesis are presented to give the reader a basic introduction.

6.9.1 Timing Paths

Synthesis tools try to ensure that the logic is implemented in such a way that all timing paths meet the required user-defined timing constraints. Referring to Figure 6-1, the main categories of timing paths are:

a. External inputs to flip-flop inputs (e.g., X to $D1$)

b. Flip-flop clocks or outputs to flip-flop inputs * (e.g., $Clk1/Q1$ to $D2$)

c. Flip-flop clocks or outputs to external outputs * (e.g., $Clk1/Q1$ to $Y3$ or $Clk2/Q2$ to $Y2$)

d. Direct input-to-output combinational paths (e.g., X to $Y1$)

* In cases b and c above, the actual timing arc starts at the clock pin but it is sometimes more intuitive to think of the path as beginning at the flip-flop output.

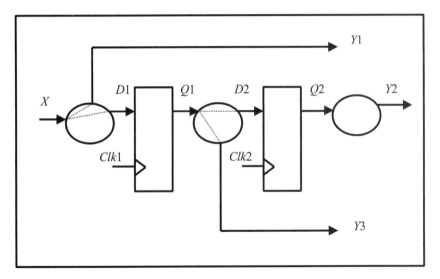

Figure 6-1 Synthesis Timing Paths

Timing constraints must, therefore, define timings from and to the extremes of these paths. For example, if an external input passes through 20 gates before reaching a flip-flop input (end of a timing path), the delay must be specified from the external input to the flip-flop input, as distinct from, for example, the o/p of the tenth gate to the flip-flop input. In fact, most tools will reject start and end points that are not true start/end points. Similarly, most tools require true start and end points as parameters to timing report commands. Artificial end points can be created in certain cases by disabling timing arcs through logic gates on a timing path, but this is not a common requirement.

Tip: If generic input delays are set greater than generic output delays in the timing constraints, be careful with the last group of paths above (i.e., direct input-to-output combinational paths). These will fail automatically because the constraints on them cannot possibly be met. This often results in the tool spending a long time trying to work against incorrectly set constraints. The solution is to reset the constraints on these paths so that they are physically realizable.

6.9.2 Latches versus Flip-Flops

From the synthesis point of view, flip-flops are preferable to latches. Flip-flops have straightforward timing paths. For example, in a *D*-type flip-flop, the *D*-input is an end point in a timing path, and the *Q*-output or the clock is effectively the start point in a timing path. However, a latch is somewhat more ambiguous. The latch input is an end point when the latch is closed but effectively a midpoint if the latch is open. Synthesis tools have special ways of handling latches, and the reader is referred to the tool manuals for details on these.

6.9.3 Timing Margins/Time Budgeting

In bottom-up synthesis, where the entire chip is not synthesized in one go, it is normal for designers to agree on timing budgets between their blocks. In such a case, true timing paths are often broken at a midpoint. For example, a combinational logic path of several gates between a *Q*-output and a *D*-input of a *D*-type flip-flop may be artificially broken in a synthesis partitioning exercise. For the sake of argument, let's say that the *Q*-output goes through a cloud of combinational logic in *block A*, terminating in the block output port, *port X*, and the rest of the path from *port X* is implemented in *block B*. In such a case, it is up to the designers of *block A* and *block B* to agree on the proportion of the total timing cycle that is available to each of them to work with. For example, in a 20-nsec cycle, they may agree to split the time available by allowing 8 nsec to *block A* and 12 nsec to *block B*. This exercise is referred to as *time budgeting*.

It is common practice in a time-budgeting approach to allow a small margin of 5–10% of the clock period to avoid any unexpected delays when the blocks are brought together. This can happen when input drive strengths and output loads are not accurately specified, resulting in a requirement for extra buffering.

It is also common practice to agree on default chipwide timing budgets where it is anticipated that nearly all block interface signals will comply with these budgets. Any exceptions to the budget must be agreed on between the relevant block owners. In large chips with tight timing requirements, it can be a good idea to agree that in as far as is practical, all outputs are registered. This usually makes it easier to meet timing budgets. For example, if block outputs are expected to settle in say the first 10% of the available time, this allows the subsequent block inputs which are driven by these outputs to have almost complete cycles to traverse any required combinational logic to the true end-point.

Finally, it can be helpful to have a requirement that all module owners present a maximum of one load to each input that drives multiple modules. This can be achieved for example by forcing each module to buffer the input signal before distributing it further to several different circuits. With this approach, the designer of the module sourcing the driving signal can know in advance what the maximum output load on the signal will be, and can build this load into the synthesis constraints for the module.

6.9.4 Characterize and Compile

This is a useful technique to improve on a synthesis that started with manual time budgeting. An example of its usage is as follows. Assume that a design consists of blocks A, B, C and D. If blocks A, B and C are synthesized separately from block D, the owner of block D will have agreed a time budget with the other blocks and synthesized accordingly. If block D proves difficult to synthesize, due to timing reasons, perhaps because the margin in the timing budget was critical, it may be advantageous to "characterize" block D and resynthesize it when all the blocks are pulled together. The characterization then looks at the realistic input strengths and output loads on block D and the true available time budget. These characteristics can be written out in the form of a constraints file that is used to resynthesize block D more accurately. Although time-consuming, this often results in improvements where the time budgeting is marginal.

6.9.5 Overconstraining Designs

It is normal to allow some safety margin when setting constraints and operating conditions for synthesis. There are several reasons for this and several ways of doing it. A common reason in large chips is that interconnect delay can be unpredictably large on a number of nets, and a design that passed all the synthesis timing requirements before layout can fail to meet certain timing paths after layout. A further reason in bottom-up synthesis using time budgeting is to allow margins for any extra buffering required between blocks, due to inaccurately specified interconnecting signal characteristics.

There are several common methods of overconstraining synthesis. The synthesis clock(s) can be defined at a higher frequency than their actual frequency. This forces the tool to try to meet the timing by ensuring that critical paths are made shorter. In real life, the clock period is longer than the synthesis period and, thus, we have our margin. Another common approach is to set worst-case operating environments that are more pessimistic than the actual expected environments. In the manual time-budget approach, the design team can choose a safety margin between corresponding input and output timing budgets.

There is no precise percentage for overconstraining that is appropriate in all cases. In fact, in some cases, no overconstraining is appropriate. Some companies use a margin of 5–10%. However, it is best to review the synthesis approach with both the ASIC vendor and synthesis tool provider, bearing in mind that the ASIC vendor's models may have built-in margins.

The up side of overconstraining is that there is less chance of timing violations when the chips are manufactured. There are several down sides. In the extreme case, it may not be possible to meet the artificially tight timing constraints, and the chip may not be feasible or may be feasible only if a more expensive, faster technology library is chosen. If the timing requirements are tight, overconstraining will result in a larger gate count as the synthesis tool attempts to minimize logic depth by making more parallel structures. The resultant larger area may result in a requirement to increase the die size. It will also normally result in higher power consumption.

6.9.6 Grouping and Weighting

If it proves difficult to meet the timing requirements in a given design, improvements can sometimes be made by using "grouping" and "weighting" commands. Grouping involves defining a set of signal paths under a group name. The group can then be given a weighting in the synthesis process. By increasing the weighting, the violation on a signal or group of signals appears worse to the synthesis engine than it actually is. The synthesis tool correspondingly gives the signal or groups of signals more attention in the optimization algorithms; therefore, there is an increased chance that the signals in the group will meet their timing requirements. This technique is useful when a small number of signals fail to meet their timing requirements but the majority of signals in related areas of the circuit have an average comfortable degree of margin. However, if all related circuit signals have marginal timing, the situation becomes analogous to a bulging balloon. By weighting and fixing one "bulge," a new "bulge" appears elsewhere. That is to say that we end up with an alternative set of violations. It is probably worth examining whether any critical paths can be improved by rewriting the source code before starting to experiment with grouping and weighting.

6.9.7 Flattening

Synthesis tools usually have a variable to determine whether hierarchies are maintained or "flattened." By breaking hierarchies, there are more opportunities for optimization. This should result in smaller areas and better performance. The down side of flattening is that it becomes virtually impossible to understand the gate-level netlist in a flattened design of any significant size. This is relevant if debugging is required at the gate level. It is generally advisable, therefore, to avoid flattening where possible. If it does prove necessary, it should be introduced only in critical sections, rather than flattening the entire chip.

6.9.8 *dont_touch* Attributes

This is a Synopsys Design Compiler term but the concept exists under different names in other synthesis tools. It refers to an attribute that is used to indicate to the synthesis tool that it should not attempt any synthesis or optimization on cells or modules that are marked with this attribute. It is sometimes used for precompiled blocks, where there would be a considerable and unnecessary time loss if the precompiled blocks were resynthesized. It is also sometimes used where certain blocks required special treatment and the use of special synthesis tricks to achieve the necessary results during their original synthesis. In such a case, a higher-level resynthesis might undo the effects of the special attention that was given previously in the original synthesis.

6.9.9 Black Boxing

If a model for a module or a macro is not available during synthesis or is removed for some reason, the synthesis can continue by treating the module as a black box. An example of this might be a synthesis run requiring a not-yet-available memory cell. Rather than hold up the entire synthesis process, it makes sense to proceed without the memory cell and effectively estimate its effect. All inputs to the black box component that are derived from the rest of the circuit are treated as outputs from the rest of the circuit. For this reason, they are defined as outputs of the rest of the circuit and are given specific timing requirements, based on estimates of the timing requirements of the missing model. Similarly, all outputs from the black box that are fed back into the rest of the chip are defined as inputs to the rest of the chip and assigned input timing constraints. The synthesis can then proceed, using these estimates until the real model is available, at which point, an incremental synthesis should be adequate if the timing requirements and characteristics of the black box model were reasonably accurate.

6.10 Managing Multiple Clock Domains

Designers often lose sleep over this topic! There are numerous steps that can be taken in design and synthesis to ease the problem a little.

6.10.1 Isolate the Asynchronous Interfaces

It is worth considering having special asynchronous interface modules for all signals that cross-clock boundaries. In this way, they can be synthesized separately and given the required attention they need, and there is no need for any tiring searches throughout the netlist for other boundary-crossing paths because they are all contained in the designated special blocks.

6.10.2 Identify Synchronizing Flip-Flops with Unique Names

Where asynchronous boundaries are managed by inserting synchronizing flip-flops, it is advisable to name the corresponding signals in such a way that the flip-flops can be easily identified. For example, adding an _sync string to a signal name will result in the corresponding register having the same string in its name. The advantage of this is that the paths can easily be identified and ignored in timing violations, or, more correctly, set as false paths, so that they no longer appear in the violations list.

6.10.3 Set False Paths Across Known Asynchronous Boundaries

For other asynchronous interface-handling methods, where the safe handling of the interface is guaranteed by design (for example, asynchronous data accompanied by synchronous handshake envelope signals), the corresponding timing paths should be set as false paths in the synthesis constraints. By doing this, we can establish that any timing violations in timing reports are genuine, and we do not have to waste time establishing which violations are true and which are due to incorrectly specified asynchronous paths.

6.11 Managing Late Changes to the Netlist

The final system simulations often reveal minor problems with the design, requiring minor changes to the netlist. Thus, it is not uncommon to hear a design team talking about not only a "final" netlist, but also a "final final netlist," and sometimes a "final final final netlist." So how are late changes best managed?

Depending on the size of the netlist, the processing power available and the degree to which the design is overconstrained, synthesis of an entire chip can take from several hours to several days. Clearly, it is a considerable setback to lose a day or several days each time a minor change is required. There are several ways of approaching the problem.

6.11.1 Complete Resynthesis

This is the most time-consuming option but possibly the safest. Once the behavioral code changes have been made and simulated, an entire resynthesis is carried out. If there are many changes, this may be the only realistic option. The advantage is that there is a clear, direct, unambiguous, one-to-one relationship between the behavioral netlist and the gate-level netlist because the gate-level netlist is derived automatically from the behavioral netlist.

6.11.2 Partial Resynthesis

This approach is useful if there are numerous changes but they are confined to one or two modules. A possible approach here is to synthesize only those blocks that have changed, possibly using a "characterize-compile script" approach on the latest gate-level netlist to generate accurate constraints. Once the individual blocks have been synthesized, the older version of these blocks can be deleted from the gate-level netlist and the new ones read in to replace them. A final top-level synthesis is then required to ensure that there are no violations as a result of knock-on effects from the changes in the modified blocks. It is possible to preserve critically timed blocks in this process by applying *dont_touch* attributes.

There are many different possible "partial resynthesis" approaches. As with complete resynthesis, there is a one-to-one relationship between the latest behavioral code base and the synthesized gate-level netlist.

Unfortunately, it is not always as simple as it might seem at first glance. Customized scripts are required to run the necessary sequence of steps. If scan chains have been inserted on the original netlist, it is tricky and often not worthwhile trying to patch something new back into the netlist. However, if scan chains were added late in the synthesis process or as a last step, it may be possible to retrieve the prescan gate-level netlist as the starting point for the partial resynthesis. The scanchains are then reinserted as a final step.

6.11.3 Editing the Gate-Level Netlist

The advantages of this approach are that it normally results in the fastest turnaround time, and there are no problems with scanchains. The big disadvantage of this approach is that the possibility of introducing errors is greater. The behavioral netlist changes and the gate-level netlist changes are made in independent processes. The gate-level netlist is no longer automatically derived from the behavioral netlist. The translation is manual. The risk associated with editing the gate-level netlist is reduced or eliminated if a formal verification tool can be used to compare the functionality of the RTL description and the gate-level netlist.

Two possible approaches to editing the gate-level netlist are described below.

6.11.3.1 Editing the Netlist Using a Standard Text Editor

This approach involves using a standard text editor to modify the gate-level Verilog. New nets and new gates can be declared and instantiated. Existing connections can be changed. This is simply an exercise in modifying Verilog code but with the constraint that any modifications must not contain any behavioral constructs. The disadvantage of this approach is that it is extremely manual and not easily reproducible or easily reviewed. The behavioral code should always also be updated when this approach is used, even though it may not subsequently be synthesized. It is still useful for behavioral simulation, and it avoids endless confusion 6 months later, when someone is trying to match up the signed-off gate-level netlist and the latest behavioral netlist.

6.11.3.2 Editing the Netlist Using Synthesis Script Commands

This is a similar approach but is more reproducible and more easily reviewed. Most synthesis tools have commands that support netlist editing. Using pseudocode examples, these are typically of the form *create_net*, *create_gate*, *delete_net*, *delete_gate*, etc. The necessary commands can be sequenced in a script that reads in the existing netlist, modifies the netlist as required and saves out the new netlist. As with the text-editor approach, the behavioral code should always also be updated when this approach is used. The same arguments apply.

6.12 Summary

This chapter has covered many of the important aspects of chip synthesis. Top-down versus bottom-up approaches have been discussed. The limitations of traditional synthesis techniques in dealing with deep submicron designs have been highlighted, and the consequent arrival of new physical synthesis tools has been noted. The merits of scripting and GUIs were presented. The script sequences and sample scripts provided can serve as a basic framework for real-life applications. Key synthesis terms and concepts, such as time budgeting, characterize-compile, flattening and black boxing have been explained. Issues such as managing asynchronous interfaces and handling special cells have been addressed. As with many aspects of ASIC design, there is very often no one narrow, correct way of approaching a problem or task. This chapter has proposed a number of ideas and approaches. For more complete coverage, the reader is referred to the synthesis tool provider's manuals and application notes.

Quality Framework

7.1 Introduction

Each project should have a project framework document that defines where all the relevant project information is stored. This document should also list the main electronic design automation (EDA) and software development tools that are used on the project and should record the version number of each of these.

The framework document should be referred to by all project team members to ensure a consistent approach to the design. It enables team members to find out quickly whether a specific document exists and where it is stored. It also facilitates archiving the project at a later stage because it details where all the relevant information that needs to be archived can be found.

The specific structure and content of a project framework document will vary from company to company, and there is no one correct optimal format. However, the following sections describe some of the items that could be included. The framework document should be updated and reviewed on a periodic basis. The current trend is to create an intranet page to act as the framework document with hyperlinks to other relevant documents or information.

The final part of this chapter describes standard company procedures and a quality framework called the *Capability Maturity Model.*

7.2 The Directory Structure

The VHDL/Verilog directory structure should be defined before any coding is done. A good structure will encourage a consistent quality approach to the design. If the directory structure is more or less standard across all projects, project team members will automatically know where to look to find specific types of files. As a general rule, it is preferable to use links to central files rather than making local copies. This reduces the risk of problems associated with referencing out-of-date information. The sample directory structure of Figure 7-1 shows a number of top-level directories.

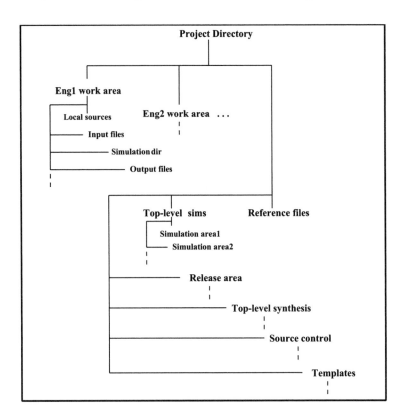

Figure 7-1   Sample Directory Structure

7.2.1 *engineer_work_area*

This is the directory tree in which the engineers create and test the design modules. Two common approaches are often seen in organizing this type of directory. In the first, each engineer has his or her own user directory with groups of modules underneath this; in the second, each top-level module has a separate directory. This decision will be dependent on the chip architecture and the structure of the team. It is recommended that the top-level directories have further levels of hierarchy to separate different files, rather than having all files lumped in together. For example, there could be a sources directory, an input files directory, an output files directory and so on. This makes it easier to find different types of files or, for example, to delete output files if disk space is running low.

7.2.2 reference_files

Module and top-level simulations often need input reference data and expected output reference files. The generation of these files requires careful consideration because they should exercise as much of the ASIC as possible. By keeping them in dedicated directories, they are easily accessible to all team members. Although all directories should ideally have some sort of README file, summarizing the contents of the directory, it can be particularly useful to have an up-to-date README file in this type of directory, highlighting the main characteristics of each of the reference files. It is also helpful if reference files are named in such a way as to give some idea of the nature of the data content, e.g., random_data_sequence_1ms.dat.

7.2.3 top_level_simulations

It is useful to arrange the top-level simulation suite under one directory, with sub-directories for the different ASIC functions or features. This approach can be very useful when reusing ASIC functions or running regression tests following minor revisions to the chip. It avoids the time-consuming exercise of hunting through each of the different individual designers' user directories to try to find the necessary set of tests to verify the design changes. It is clearly a much simpler exercise if all top-level simulation tests are located under one directory tree.

7.2.4 release_area

The latest stable top-level register-transfer-level (RTL) description and gate-level netlists should be checked out to a central release directory, and all top-level simulations should link to this area. The top-level simulations should be run on a specific version of the netlist, and when the results of running suites of tests are being documented, the specific version of the netlist that was used should be recorded. Updating the release_area directory with new netlists should be done in a controlled manner. When the directory is updated, all members of the simulation team should be notified to ensure that any necessary simulations are rerun on the final netlist.

7.2.5 Source Control

A source code control system should be used to store all source code files. In the case of ASCII files, source code control systems can store multiple evolving versions of the same source document very efficiently. For example, it can store the changes between two versions, rather than the entire new file each time. The source code control directory tree can be divided into different subdirectories to store different types of sources.

7.2.6 Synthesis Directory

The synthesis directory is used to store all synthesis-related files. Most obvious among these are the synthesis scripts and the synthesized netlist output files. As with other directories, it is useful to create subdirectories in the synthesis area, also. For example, there could be separate subdirectories for scripts, constraint files, output netlist files, synthesis reports, back-annotated timing data, etc. Depending on the preferred directory layout style of the project, the synthesis directory may also be the area where the top-level hierarchy of the chip RTL description is created and stored. This is not normally stored under an individual user's directory, and the synthesis directory is often a logical place to create it.

When generating and storing new synthesized netlists, it is important to have a sensible naming convention on each new release of the netlist that allows it to be easily identified and retrieved. When sending off netlists to the ASIC vendor, it is essential to keep an accurate record of which netlist was sent and the key characteristics of that netlist that differentiated it from the previous one. One common naming convention is to add a date suffix to the netlist name, e.g., *full_chip_02jan.v* or, alternatively, to create a separate subdirectory (again, with a date suffix as part of the name) for each release of the gate-level netlist. Ideally, each netlist should have its own specific associated README file containing a summary of the main changes in that netlist, compared with the previous one.

7.2.7 Templates

There is a number of common scripts that can be useful for all the project team members, such as general-purpose synthesis scripts, makefiles, sample control files, etc. This directory should contain all of these useful scripts. Note that the main design-specific synthesis scripts are stored somewhere in the synthesis directory tree and that the synthesis scripts referred to in this example are general-purpose templates or common utility scripts. Templates for module specifications, design implementation notes and any other document templates should also be made available in a central directory.

7.3 Documentation Storage

Documents and other information should be stored on a central server, allowing all project team members access to the files relevant to their needs. Care should be taken in granting access rights to sensitive files. All project documents should be listed in a file or database (which is identified in the framework document) to allow easy searching for a particular document. Such a list should contain the file names of the documents, their location, the author, the creation date and a description field or description identifier type. If the list is in the form of a spreadsheet or database, or even an MS Word table, the description type/author/date fields, etc., can be used to sort the information. Each project document should have a unique project reference number that is recorded in the spreadsheet/database.

7.4 Freezing Documents and Controlled Updates

Some important documents, such as the requirements specification, should be reviewed and formally approved or "signed-off." Once the document has been approved, there should be a documented procedure for any updates. This means that the document cannot be changed unless formally approved by a group of nominated representatives. All documents that are treated in this manner should be defined in the project framework document. It is useful to have a table in the project framework document that defines all controlled information, when it is frozen, who the nominated representatives are, and the current approved version. A standard version numbering system for documentation can be useful.

7.5 Fault Report Database

Bugs should be recorded and tracked in a central database. All engineers should have access to the database so that all bugs identified can be formally recorded in one central location. The bugs list should be reviewed at least once per week at the project meeting. Bugs should be prioritized and assigned owners. The rate of creation of bugs, compared with the rate of resolution of bugs, gives a very useful indication of the stability of the netlist during the top-level simulation phase.

Bug recording and change requests are frequently stored in the same database. All changes to frozen documents, such as specifications and plans, should be recorded in such a database. A centralized database that records all change requests provides a structured way to track, plan and execute tasks that fall outside the original project plan.

7.6 Source Code Control

Source code control is an essential requirement for any project hoping to achieve quality results. Once a project reaches a certain critical size, the absence of source code control is a guaranteed recipe for chaos. Source code to be archived should include synthesis and other utility scripts, as well as the more obvious design module code. Other files, such as requirement specifications or project plans, should also be source code controlled. This can be useful when tracking differences in the specifications or approved project milestones and deliverables.

Enforcing source code control on design code from day one is not practical because it can lead to hundreds of versions of each of the design files being stored, with a consequent negative impact on overall storage space in the case of large projects. Experience usually dictates when it is best to initiate source code control. As a guideline, the code should have some degree of stability and, preferably, should have been reviewed and tested to some extent. The real benefit of source code control is the ability to retrieve known working versions of the code when updates suddenly cause previously passing tests to start failing. When carefully chosen, meaningful comments are added with each version of the code registered in the source code control repository, it helps to narrow down the reasons for unanticipated behavior.

7.7 Makefiles/Simulation Scripts

Toward the end of a project, the focus on quality can easily disintegrate when time pressures tempt team members to ignore certain procedures and cause team communications to deteriorate. Makefiles and simulation scripts can help to preserve the good practices executed throughout the project. Makefiles help to ensure that all users are referencing the latest released files. Simulation scripts that invoke a suite of regression tests can be used to verify that late changes have not caused a problem with simulations that were previously working.

7.8 Company-Defined Procedures

Many companies define the process of designing ASICs within documented procedures. The quality of these procedures varies from company to company. When the quality is high, the procedures provide a useful guide to the project team members, and they create a framework that allows senior management to track the progress of the project carefully. With good procedures, the organization can plan and execute projects very efficiently, even when working with new technology or with engineers new to the team. When the quality of the procedures is low, they add irrelevant extra administration tasks to the project or they are simply ignored by the team.

Good procedures can improve efficiency by incorporating best-of-class techniques into the project. However, there are many areas that procedures can address, and the most difficult decision for companies is which procedures to focus on first. The Carnegie Mellon University Software Engineering Institute has developed a framework that defines the maturity of company processes and procedures. This is called the *Capability Maturity Model* (CMM). The goal of the CMM is to develop an organization's people, process and technology to improve long-term business performance. One of the most useful features of the CMM is that it shows how a company can evolve its processes and procedures over a number of years so that projects deliver the right goals at the right time. The CMM is focused on software development, but many of the ideas are applicable to ASIC design.

The CMM is defined in CMU/SEI-93-TR-25 *Key Practices of the Capability Maturity Model*, version 1.1, by Carnegie Mellon University. CMM information is also available via the CMU web site.

7.9 Summary

This chapter describes a quality framework that ensures that information is stored centrally and that updates are implemented in a controlled manner. It suggests a common directory structure for the VHDL/Verilog design environment and the use of source code control tools to store archive versions of the code. Change requests and bugs in the design should be tracked using a central database. Company procedures can be a useful guide to the project, and the CMM can help organizations to develop detailed procedures.

## CHAPTER 8

Planning and Tracking
ASIC Projects

8.1 Overview

Planning and tracking form core roles of the project manager. However, these are not simple tasks. Like most tasks, the best way to obtain a good result is to base the plan on previous plans and the experience of other project managers. Approaching the task in a systematic manner will make the experience less stressful and produce more positive results. Tracking and monitoring progress is key to ensuring that the project goals are achieved in a timely manner. Keeping the plan up to date is very important for accurate tracking.

The first part of this chapter identifies some basic planning concepts. The second part deals with creating the plan, analyzing the plan and optimizing particular areas of the plan. The final part of the chapter deals with tracking.

There are several institutes that specialize in project management disciplines. These include the APM (Association of Project Managers), the IPM (International Project Managers), and the PMI (Project Managers Institute). Each of these institutes is a source of further detailed information in this area.

8.2 Basic Planning Concepts

There are some important concepts that should be understood before attempting to produce an ASIC development plan. First, it is important for all project team members to understand the plan and the reason for the plan. There are many objectives behind planning. Principally, the plan should identify all the required tasks, the sequence in which these tasks should be done and who should be assigned to carry out each task.

The second basic concept to understand is that the level of detail in the planning changes at various stages in the project, becoming more detailed as the project progresses. There are three significant milestones at which the level of planning detail can be significantly improved upon. It is useful to view these almost as separate plans because the level of detail at each stage effectively takes a quantum leap.

- An initial high-level plan is needed for the feasibility business case.
- A second, more detailed plan can be generated after the architecture design has been done.
- A third, yet more detailed plan can be put in place after module breakdown, when all the module tasks have been identified and the level of effort has been estimated. This third plan should not be viewed as a final, cast-in-stone plan or static document—it should be updated and modified as issues and unplanned activities are identified. However, ideally, it will be sufficiently accurate from the outset not to warrant overfrequent changes.

Each of the three plans should contain an appropriate amount of detail but the latter plans will contain more detailed and more accurate information. The senior management team should be made aware that the initial plan could be inaccurate by a margin of up to 25%, depending on the scale and complexity of the ASIC.

The initial plan will be generated by the initial core team and should identify only top-level tasks. The focus of this plan is to estimate the expected sign-off date and the required effort, rather than identifying every individual task and identifying precisely who will carry out the tasks.

The project team members will be more committed to achieving the milestones and deliverables set out in the plan if they have helped to create it. Depending on how the team is phased in, the opportunity to involve the majority of the team in the planning process normally arises in the second or third planning phases referred to above.

Following the architecture design, when the full team or a large part of it is assembled, a first-pass definition of the tasks involved in each module can be generated. The focus at this stage is to identify the complexity, risk level and resource requirements associated with each module.

The third planning session takes place following the module design breakdown, when the full complexity of the lower modules can be assessed with a reasonable degree of accuracy. Until this breakdown has been done, the tasks cannot be accurately estimated.

The third basic concept of planning is to understand the relationships and trade-offs between three fundamental components that underline the plan: the functions/features to be designed, the end date and the development/product cost.

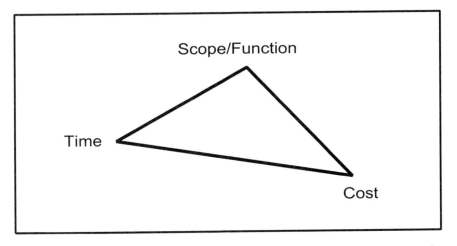

Figure 8-1   Cost/Time/Function Planning Triangle

Figure 8-1 shows a classic project management triangle that defines these three components of the plan. One corner represents the functions or scope, one the end date, and the third the cost. The triangle is drawn to show that each of these functions is related. If one corner of the triangle is changed, the characteristics of at least one other corner must change. Time to market can be reduced if the cost is allowed to increase or the number of functions or features is reduced. This is typical of projects that require a new generation of ASIC, for example, when designing a product for a brand new market area where time to market is critical. However, for cost reduction projects, the cost of development can be more critical than time to market, and the end date can, therefore, be relaxed.

The fourth basic concept of planning is to understand that some projects are significantly different from previous projects and, therefore, require a new approach to the design. There is a general industry-wide trend toward the development of larger, more complex ASICs. As design size exceeds certain thresholds, it is usually necessary to make some significant changes to the design approach and the associated planning. Many companies follow a similar phased approach to ASIC design, starting out with smaller designs and gradually evolving to take on larger, more complicated developments.

When a company decides to start up an ASIC design activity, the first ASIC is normally relatively small (less than 100-K gates being small at the time of writing).

However, because the design flow and techniques are new, there is considerable risk with the initial timescales. Ideally, therefore, the first ASIC should not be on the critical path of a very important project.

After designing a number of smaller ASICs and as in-house experience grows, the timescales can be defined with a greater degree of confidence, and the risk of major slips to the end date should be relatively low. The size and or complexity of the ASICs typically tend to grow at this stage.

The risk of the end date slipping will stay low until a size or complexity threshold is passed (e.g., over 100-K gates). This threshold requires new techniques to ensure that the end date is achieved on target. The design team normally grows in size, and the management of the team becomes more important. There are many changes that can be implemented to help smooth the transition to larger ASICs.

- The organization of the team can change, creating roles for specialized team members, e.g., synthesis or design-for-test (DFT) experts.
- Common approaches to design can be used, e.g., a specific coding style or methodology.
- New tools can help the transition.
- Reuse methodologies can be adopted to limit the new design work.

After further size or complexity increases, yet more thresholds are passed. These will require more changes to the design techniques and processes. Additional tools may be required (e.g., formal verification or floor-planning tools), computing resources may need to be upgraded and the design reuse approach may need improvement. Adoption of new tools normally requires additional skilled engineering team members to manage the tools and get the maximum benefit from their usage.

A key task for the project manager in all of this is to be able to identify these critical size/complexity thresholds and to adapt the planning approach accordingly. When previous projects have shown some signs of strain (e.g., top-level simulations become very slow), this is normally a sign that the next project requires a fresh planning approach and that a critical threshold is being crossed.

8.3 Process for Creating a Plan

Creating a plan can be viewed as a six-step process:

- definition of deliverables
- task breakdown

- assignment of dependencies
- allocation of resources
- refinement of the plan
- review of the plan

These steps are discussed in detail in the following sections. By way of introduction, the remainder of this section first gives a summary explanation of each step.

Defining the deliverables involves identifying and clearly documenting all outputs of the project. Task breakdown involves identifying all the tasks needed to complete the project. For ASIC projects, this step cannot be accurately completed until the top-level architecture and at least some of the lower-level architecture is designed. Some tasks cannot begin before another task has completed. The "assigning dependencies" step identifies those tasks that are dependent on the completion of another task. Resource allocation associates specific resources with specific tasks. The resources referred to include nonhuman resource, such as special machine usage, development hardware, etc., in addition to direct human resource. After the initial plan has been created, the team should then spend some concentrated effort on refining and improving the plan, generating ideas to reduce risks and, if possible, bringing in the end date.

The following subsections describe these steps as they apply to ASIC development projects.

8.3.1 Definition of Deliverables

The features and functions of the ASIC must be clearly captured in some form of written specification, often referred to as a *functional requirements specification*. This should be done in consultation with marketing engineers. Each feature and function can be categorized as either a "must-have" core requirement or a "wish." Items on the wish list can be removed if necessary, for example, if they have a disproportionate negative impact on time to market. It is important that the features and functions are well defined. Otherwise, there is a risk that the plan will be inaccurate from the start. Marketing and senior management should formally agree on the functional requirements specification.

From the features and functions list, an architecture is designed that is capable of satisfying all the requirements. This identifies the main functional building blocks in the ASIC and forms the basis of the plan. At this stage, it is important to identify any blocks or subblocks for which already existing modules can be reused. This reduces timescales and risk. The potential for future reuse of all or parts of the new design should be identified, because this, too, can affect the architecture and will certainly increase the time needed to complete the task.

8.3.2 Task Breakdown

It is useful to subdivide task breakdown into two categories: design module-specific tasks and other, more general tasks.

Consider first the design module-specific tasks. These include all tasks associated with designing, testing and synthesizing each module or function in the chip. Identifying the tasks is best done in a top-down methodology. Starting from the top-level architecture, each module or block should be decomposed into smaller and smaller modules or subblocks. Spending time at this stage will improve the definition of each task and results in more accurate estimates of task durations. Chapter 3, "A Quality Design Approach," covers a quality approach to module design that can be used to help identify module tasks.

Having broken down the top-level architecture into a series of low-level tasks, each task is then analyzed in terms of estimated effort, duration and risk. A task might be defined as high risk due to the complexity of the design or because of the impact on the project if the task runs late. For example, a CPU interface may be relatively simple but, typically, it must be fully functioning before any top-level simulations can start. A delay on such a task could delay the entire team and the entire project if the task is planned to finish as the top-level simulations start. High-risk modules should be designed early in the project. This allows replanning if the module takes longer than expected to develop. At this stage, all predictable risks should be considered and where appropriate, extra tasks should be identified to reduce project risks (see Chapter 9, "Reducing Project Risks").

The estimate of effort is a complex task that should be based on input from a number of team members. If historical planning data on similar modules designed in previous projects exists, this is a good starting point. For the initial estimates, the tasks should be given an effort or duration figure based on the work rate of an average engineer with three or four years of design experience. The actual effort can then be updated after the specific resource has been allocated, being scaled down for more senior engineers and being scaled up for less experienced engineers.

A useful technique called *expert judgment* can help with the estimation. A few senior engineers and the module designer independently estimate the task. The estimates are compared, and the final decision is based on the amalgamation of the individual calculations. If one estimate is significantly different from the others, the reason for the variation should be discussed.

The estimated effort or duration should be checked against another metric, such as the estimated number of lines of code or estimated size of the module in silicon gates in conjunction with the complexity of the module. A historical average number of lines of code per engineer (e.g., 30–100 lines per week) or silicon gates (e.g., 100–500 gates per week) can be calculated from previous project data. This figure will be highly dependent on the type of ASIC designed and to an extent on the development tools used. Using this metric and the estimated size of the module to be designed, a required effort for the task can be estimated. This should be cross-compared with the estimate based on expert judgment. Creating obviously unrealistic timescales to get team members to work harder does not work. If the timescales are unrealistic, the team will ignore them and become demotivated.

The second category of tasks referred to at the start of this section includes the additional non-module-specific tasks that are necessary for the completion of the ASIC design. These tasks are generally relevant and common to all chip developments. They include creation of common synthesis scripts, creation of the top-level netlist, synthesis of the top level, system testbench work, system and subsystem simulations, test insertion, generation of test vectors, formal verification, etc. Because they are common to all projects, these tasks can be defined in a template that is available for all projects.

Following task breakdown, a list of all predictable tasks will be available with initial estimates for effort required and some assessment of the associated levels of risk.

8.3.3 Assigning Dependencies

The next step is to add dependencies to the tasks. This means defining which tasks must be completed before other tasks can begin. In some cases, these dependencies are obvious, such as the fact that a module cannot be tested before the code has been written. In other cases, the dependency is more subtle. For example, it may not be possible to design a processing block fully until a statistics block requiring specific status signals from the processing block has been fully defined.

Once all tasks have been identified, complete with initial estimated durations and associated task dependencies, the theoretical sign-off date can be calculated, based on initial estimates of available resource. This provides the starting point to the realistic sequencing of each task.

This sign-off estimate is done easily, using a project planning tool such as the MS Project with the feature called *leveling*. The project-planning tool will require that each task be given a name, estimated duration, dependencies and resource name. To establish the theoretical earliest sign-off date, the resource for all tasks can initially be set to a defined name, such as *Resource1*. MS Project has a resource sheet in which the availability of each resource can be defined. For this exercise, the resource *Resource1* should be set to model the total number of engineers in the team (e.g., a five-man team would be 500%). Leveling can now be done, which results in the theoretical sign-off date. At this point, other than the dependencies defined above, the sequencing of the tasks will be random. However, at this stage, this is not important.

It is useful to experiment with the project planning tool, varying the available total resources to assess the impact of extra or reduced resources on the end date, because this is a question often asked by senior management.

8.3.4 Allocation of Resources

With the theoretical sign-off date established, tasks can be assigned to specific team members. A number of factors should be considered when assigning the resource. There are project-related considerations, such as matching the complexity and risk associated with the task with the knowledge and experience of the team members and the actual availability of the resources (often, senior engineers will not be available 100% of the time). Although project considerations are important, there are other considerations, too. Is a specific task appropriate to the development path of a particular team member? Is it sufficiently challenging or interesting? To satisfy the development needs of the individual team members and to maintain team morale by providing each person with a sufficiently challenging and interesting work load, the allocation of resources to tasks becomes a complex juggling act.

Once the initial allocations have been made, the standard durations of the tasks (assigned in the previous step) should be reviewed and updated, where necessary. Typically, senior engineers will complete tasks faster than the standard, and less experienced engineers will take longer. It is useful to have a structured approach to assigning the resources, starting with the highest-risk modules and ending with those tasks needed to develop the least important features.

Once the resource allocation has been done, the scheduling of the tasks can be started. Writing the task and duration on a Post-it note, then sticking these on a wall can speed up this activity. This makes it easy to experiment with different sequences and see the effects that each sequence has on key milestones and project end date. After some trial-and-error experimentation, the task sequence should be entered into a project-planning tool so that an accurate analysis can be carried out.

Because this is the stage in the planning process where the plan begins to take on its final form, it is worth considering the following tips when entering the plan into the project-planning tool:

- Include holidays in the plan, both general holidays and the holidays of the individual team members.

- Ensure that the plan has some contingency. This could be some resource that is not fully used throughout the period of the project. Some project managers include a task called *contingency* as a means of including contingency.

- Avoid having all team members on the critical path. If anyone slips, the end date will slip. Having just one or two engineers on the critical path lowers risk and allows the project manager to focus on the critical path.

- Allow time for design reviews, bearing in mind that some of the more senior engineers will attend many reviews as members of the reviewing panel, in addition to the time required to present their own work for review.

- If there are many inexperienced members on the team, allow some time for the senior engineers to provide coaching support to these.

- Allow learning curves for users of new tools or for users new to or unfamiliar with existing tools.

- Ensure that the team receives adequate training at the appropriate times and that this is entered as an activity in the plan.

- Make the plan readable and understandable. This is important because all team members should refer regularly to the plan throughout the project. Adding notes and summary tasks to the Gantt chart makes reading the plan easier. It is also beneficial to give the team some simple training on the planning tool, such as how to filter the tasks that are in their name, how to print specific calendar periods, etc.

8.3.5 Refining the Plan

Having identified all the tasks and allocated resources to each task, the initial plan is now in place. At this stage, some analysis and refinement of the plan can take place. The first area to focus on is the critical path to see whether the project end date can be brought in by somehow shortening the critical path. It may be possible to reduce timescales by employing some of the following techniques:

- Assign those tasks with longer duration or those on the critical path to more senior engineers. They will typically complete the same tasks as much as 30% quicker than an average engineer will.

- Consider adding extra engineers to some of the tasks on the critical path.

- Consider changing the architecture or feature list.

- Consider changing the sequence of the tasks to reduce the critical path. Sometimes, the dependencies of the tasks can delay the project completion date.

After critical path analysis, the plan should be reviewed for risks. Particular attention should be paid to the critical path tasks and those assigned to the less experienced engineers. The risks may have changed, now that the resources have been allocated. Ask the following questions in relation to each task:

- What is the likelihood of the task running late?
- What is the impact on the plan if the task runs late?
- What preventive action can be taken to reduce this risk?
- What is the worst-case scenario and what contingency actions can be taken in a worst-case scenario?

Finally, milestones should be added to the plan. It is always important to have a number of short-term milestones identified on the plan. These help to keep the team focused and motivated. The short-term milestones can be added once a month on a rolling basis, as the status of the project at that particular snapshot in time becomes clearer. Additionally, the main project milestones (such as "architecture defined," "chip integration complete," etc.) should be identified on the plan from the outset. These long-term milestones are good overall progress markers and are useful when reporting the overall project status.

8.3.6 Reviewing the Plan

Reviewing the plan with the team members and peers is a good way of improving the quality of the plan. It is useful to base the review on a checklist, so that simple questions are not overlooked:

- Are holidays included?
- Are training days included?
- Is time included for attending reviews—both presenting one's own work and attending reviews of the work of others?
- Check for overloading of resources, i.e., are people working more than 100% of the time?

- Check that all the tasks have assigned resources—if not, initiate steps to acquire more resources. It is better to highlight a lack of resources early than to report a slippage later in the project.
- Consider the experience of the team. What measures are being taken to ensure that the less experienced people have help and support? How much time will be lost to senior team members in coaching less experienced team members?
- Check that the team members are assigned full time to the project. Is there a possibility that any members of the team will be assigned, either temporarily or permanently, to other projects before this project is completed?
- Check that appropriate milestones have been defined.
- Check how deliverables to and from other groups are specified and controlled. Avoid ambiguity.
- Check that some risk reduction steps or tasks have been defined. For example, part of the design may be prototyped in an FPGA. If so, the associated tasks must be present on the plan.
- Check the holiday dates for the silicon fabrication plant.
- Check the holiday dates for the ASIC vendor center that will do the layout, and check the individual holiday dates of its key technical support people for the project.
- Analyze the critical path—can it be shortened?
- Analyze risks associated with the main tasks. What has been done to reduce the risks? What contingency is available if critical path tasks run late? Are there tasks that can be cancelled to reduce timescales or risks, without major impact on the functionality? For example, certain tasks could be done in software, rather than hardware.

8.4 Tracking

Tracking is a crucial role of the project manger. The plan is of limited value unless it is used throughout the project to monitor progress. Tracking is the process of updating the plan with completed tasks and analyzing the effect of slips or changes to the plan. Careful tracking also provides important information for future projects, leading to more accurate plans.

The plan or a separate budget sheet should also include budget information. Careful tracking will highlight potential spending overruns due to changes or slips to the plan. Future plans can also benefit from analysis of the manpower costs of previous projects.

Tracking should be carried out in a formal and an informal manner. The formal tracking can take place during a weekly project meeting. The meeting should focus on progress compared with the plan and any issues encountered by the team. After each project meeting, a number of follow-up activities should take place:

- The plan should be updated with the reported progress.
- Any newly identified tasks should be added.
- Meeting minutes, including a list of actions, should be circulated as soon as possible.

Informal tracking can be done on a daily basis by talking to individual engineers. This is particularly useful at critical times during the project. Reviews are also useful for identifying the progress on each module.

Accurate tracking is achievable only when the individual tasks have a short duration or have milestones spaced at relatively short intervals. Tasks can be split into two categories: short-term and long-term tasks. The definition of a short-term task will depend on the total duration of the project, but for a typical complex ASIC, a short-term task might be one that starts within the next two months. These tasks should have a maximum duration of the order of 5 days, or at least have interim milestones with this level of interval, so that any slippage is obvious at the weekly progress meeting. The long-term tasks can be of greater duration. Once a month, some long-term tasks become short-term tasks and, at this transition stage, these tasks should be broken into a number of shorter tasks.

Throughout the project, the project manager should pay particular attention to the high-risk and critical path tasks. Critical modules should have a number of regular associated milestones and, ideally, for each critical module, some contingency plan needs to be created. Any slips to the timescales of critical modules should be identified as early as possible. The effect on the overall plan can then be analyzed and changes put in place to reduce the impact of the slippage.

It requires a conscious effort to keep the plan up to date. Time pressures on the project manager often result in the tracking activity being ignored or attended to infrequently. This is unwise and unsafe. Time should be allocated once every week to update the plan.

8.4.1 Tracking Methods

Several tracking methods have been devised to show the progress of the project graphically. These include the slip-lead chart and the earned-value chart. The senior management team members often want to understand how the project is progressing against the plan. This can be done using the charts or simple graphs of predicted budget against actual spending, and predicted milestone dates compared with those achieved. Tools such as MS Excel can generate these graphs easily. It is useful if the company uses a standard spreadsheet template for presenting the information.

8.5 Summary

This chapter has dealt with the important topics of project planning and project tracking.

It begins by describing some planning concepts. The plan should be explained to and understood by the team. The initial plan can and should be replaced by more detailed plans at various phases of the project, first, when the architecture has been designed and again when the architecture has been broken down into lower-level design modules. Planning inevitably involves making trade-offs, with the principal trade-offs being made between cost, functionality, and time to market—all of which are highly interdependent. Although plans from previous projects provide a valuable starting point in creating new plans, it is important for the project manager to realize that there are certain ASIC size and complexity thresholds which, when crossed, may require a fresh approach to planning. The previous planning approaches may not be able to handle the newer, more challenging designs.

The second part of the chapter describes a six-step process for creating and maintaining a plan. In sequence, these steps are: defining the deliverables, breaking down the tasks, assigning dependencies, allocating resources, refining the plan, and finally reviewing the plan.

The final section of the chapter highlights the importance of accurate tracking of the activities on the plan. Effective tracking requires that longer tasks be broken down into a series of shorter-term tasks or, alternatively, that slightly longer tasks have clearly defined milestones occurring at short, regular intervals. This makes it much easier to assess progress against the plan. Progress tracking should take place formally at weekly team meetings and less formally when time permits, through daily conversations with individual project team members.

## CHAPTER 9

Reducing Project Risks

9.1 Introduction

One of the most important roles of the project manger is to manage risk. All ASIC projects inevitably entail a degree of risk. In many cases, risk taking is necessary to run a successful project and gain a competitive advantage. For example, there may be a choice between a low-risk but longer-duration development path and a medium- or high-risk path that would result in the product reaching the market considerably earlier, if successful. The project manager's skill lies in identifying the risks associated with the options available, then judging the appropriate amount of risk to take, given the circumstances that apply in a particular situation. Having opted for a particular risk choice, contingency plans should be developed that can be put into place if the outcomes of the risk activities do not produce the expected results. Risk management also involves taking steps to minimize and monitor the known risks. Outcomes of particular actions should be constantly reviewed and, if necessary, contingency steps put in place in a timely manner.

An effective method of reducing risks is to adopt a quality approach in all phases and aspects of the design (see Chapter 3, "A Quality Design Approach"). The project leader should promote a quality culture in the team and lead by example in this regard.

9.2 Trade-Offs Between Functionality, Performance, Cost and Timescales

Much of the risk associated with a project is determined by the early decisions taken during the initial project definition phase, where the trade-offs between functionality, performance, cost and time to market are made. To help make these trade-offs, it is important to identify and prioritize these criteria. For example, if time to market is the most critical factor, it may be necessary to limit the functionality or performance. Trying to achieve the highest level of functionality and performance while simultaneously trying to achieve short development timescales inevitably results in higher risk. In certain cases, this may be an appropriate course of action to take. In other cases, it is not. It is important to realize that, when such conflicts exist and trade-offs have to be made, it is generally true that the less trading-off that is done, the higher the consequent risk will be. Trying to achieve everything at the same time is a high-risk approach.

To take another example, if product cost is the critical factor, one way of reducing cost may be to increase the level of integration in the ASIC and to incorporate functionality that would otherwise be implemented in external peripheral circuitry. However, the greater the level of functionality in the ASIC, the greater will be the risk of bugs in the ASIC.

The best trade-off to make is not necessarily the one that involves the lowest amount of risk. So how should decisions be made at this stage? A common and quite effective approach is to consider several options with different levels of trade-offs between the key criteria. Each option is then analyzed in terms of projected development cost, product feature set, product performance, development timescales and the consequent development risk associated with the chosen trade-offs between these criteria. From the marketing point of view, the trade-offs must be assessed in terms of the impact on target market segment and potential market share.

Making the correct decisions at this initial project definition phase is crucial to the success of the project. It requires good leadership skills and the input of experienced technical and marketing people. What are the common dangers?

- As a general rule, marketing people will often want all possible features combined with a high performance while not realizing that this cannot be achieved in the necessary timescales or, if it is nearly achievable, they may not appreciate the risks involved and the consequent higher likelihood of failure. Experienced marketing people are less likely to fall into this trap and will recognize the need for trade-offs; they can help in evaluating trade-off decisions.

- An inexperienced technical project leader may feel pressured into agreeing to unrealistic timescales to meet unrealistic marketing and senior management expectations. When the pressure builds up from the marketing team— and, in some cases, the senior management team—it takes a strong personality to be able to resist the pressure to attempt what is likely to be unrealistic. It is sometimes easier to cave in and cling to the vague hope that it can be achieved. This should be avoided at all costs. It is better to be realistic and possibly cause minor disappointment up front at this stage than to have to break significantly worse news later on in the project, after significant investment has taken place.
- Overengineering is always a source of risk in a project. Overengineering results in unnecessary design complexity. This can arise from a combination of technical and marketing interests. The marketing people want all the clever extra features. This often arises out of the need to meet "tick-box" requirements, i.e., they want the product brochure to be able to claim lists of features that are rarely ever used. This is arguably necessary to compete with rival products but, alternatively, it can be argued that it is a failure on the part of the marketing team to persuade the market that many of the competitor's features are irrelevant and unnecessary.

  Overengineering can also often stem from an overenthusiastic technical team looking for ever bigger technical challenges. There is a direct link between complexity and risk. It is, therefore, advisable to opt for the minimum level of complexity required to meet the real, rather than the imagined, needs of the product.

If the correct decisions are not made in the early project definition phase by making the correct balanced trade-offs between often-competing criteria, the project is likely, at best, to be only partially successful and, at worst, a complete failure. It is a vital decision-making time in the project and requires the cooperation of experienced people from multiple disciplines. The project manager is advised to consult widely and arrive at a balanced view, based on all inputs and previous personal experience.

9.3 Minimizing Development Risks

Once the high-level trade-offs between cost, functionality, performance and time to market have been made, the ASIC project leader must then get down to the task of running the project on a day-to-day basis and minimizing the known risks associated with each task.

9.3.1 Selecting the Team

The development team must have the correct balance of people with access to a selection of specialists for specific tasks, such as architecture, synthesis, testing, etc. It is easy to say that having an experienced team will result in the lowest level of risk. However, in reality, demand normally outstrips supply when it comes to skilled and experienced ASIC designers. The project leader must, therefore, negotiate for the best team available while shouldering some of the responsibility of developing less experienced team members. It is important that less experienced designers mentored by experienced team members and that their work is more closely monitored and reviewed.

If a full team is not available at the start of the project, due to resource shortages, the project leader should analyze the risks associated with not being able to get the missing people in place at the required time. The impact of this problem arising should be highlighted to senior management before a decision to undertake the project is made.

9.3.2 ASIC Architecture

The ASIC architecture has a large influence on the level of complexity of the design and the consequent associated risks. It is advisable to have a number of experienced architects and senior designers involved in defining the top-level architecture. Borrowing experts from other teams to contribute to the architectural design for several days—or at least to review the architectural design—usually proves worthwhile. Within limits, the more experienced the brainpower that is thrown at the problem at this stage, the better the results will be. In adopting an architecture, it is important to analyze all of the building blocks from a number of perspectives. Ask the following basic questions:

- Can an existing design block be reused to meet the required needs? If it can, it should reduce both risk and timescales, provided that it was designed with reuse in mind. If it was not designed with reuse in mind, exercise caution. It may be easier to design it from scratch again.
- How complex are the modules? Can the same result be achieved with an alternative, simpler architecture?
- How easy or difficult will it be to test?
- What is the likely gate count and power consumption?
- Does the team have the expertise to design all the required modules?
- Does the architecture allow partitioning into modules that can be tested largely independently at the module level?

9.3.3 High-Level Architectural Modeling

High-level modeling can be used to analyze the architecture of the ASIC. The benefit of this approach is that it can expose potential limitations in the architecture at an early stage, before the project has wasted weeks or months working on an architecture that subsequently proves inadequate. Without the early architecture check, the architectural weaknesses may not be exposed until the system simulation phase, at which stage, it may not be possible to work around the problems easily. If an architectural model is written at a relatively high level of abstraction, it will run very fast. This allows quick testing of the basic functions and algorithms. It also allows researchers who may not be familiar with lower-level hardware languages to develop and test their algorithms.

Architectural modeling is often done in a high-level language such as C, C++ or system C. Alternatively, Verilog or VHDL can be used if a higher-level behavioral modeling approach is adopted and low-level, hardware-based constructs are avoided. The idea is to model key aspects of the architecture without including all of the time-consuming, low-level details.

FIFO and memory sizes, bus speeds, memory controller arbitration logic, etc., can all be modeled and analyzed. External stimuli can be modeled and the performance of the architecture checked when exposed to worst-case conditions. The high-level model and the external stimulus test environment should be clearly defined and reviewed. There should be a specific brainstorming session on how to generate external inputs that would result in worst-case conditions. The testbench should include a facility for generating pseudo-random input events, because this often exposes limitations in the architecture that were not anticipated, precisely because of their random nature. Ideally, the high-level test environment should be self-checking and should automatically run all the different tests. The resulting input and output files should be used during testing of the VHDL/Verilog register-transfer-level (RTL) code.

9.3.4 Interface Specifications

The interfaces to the ASIC need special consideration and review. A major source of problems in ASICs arises as a result of misinterpretation or ambiguous specification of the interfaces between chips. Signals appear at the wrong time or are of the wrong type or duration. At the very least, the interfaces should be thoroughly reviewed. However, ideally, board-level simulation should also take place to minimize the risks of these types of problem going unnoticed. If the interfaces are being simulated, the risk of misinterpretation is further reduced if the testbench and test suite are specified and coded by someone other than the designer responsible for coding the interface itself.

The chances of two independent parties adopting the same incorrect interpretation are lower.

Note that the same rigorous attention to detail applies to intermodule interfaces internally within the chip. However, problems here are usually spotted more easily and, as a result, they generally have less catastrophic consequences.

9.3.5 Managing Changing Design Requirements

As a general guideline, do not lightly accept customer design change requirements after the architecture design. The customer may be an external third party or the internal marketing team. One of the most frustrating and demoralizing things for an ASIC development team is continually changing design requirements after the architectural design has been approved. From the risk perspective, changes can be handled either by modifying the existing architecture if the changes are relatively small, or by rearchitecting completely, if the changes are significant. Both modifying the existing architecture or rearchitecting completely involves risk that should not be undertaken lightly. Always remember that lots of small change requests (which might individually seem reasonable precisely because they are small) at some stage amount to the equivalent of one big change request that would have been refused as unreasonable.

Modifying the architecture can result in trying to get it to do something it was not really designed to do, and the risk of unforeseen side effects arises. Rearchitecting completely is less risky in terms of design failure but, clearly, it significantly affects the development timescales and, thus, the risk of not hitting the market on time. It is less obvious but nonetheless equally true that a sufficient number of small changes also significantly affect timescales. Remember that 20 small changes that require only one man-day each result in four extra man-weeks of work. However, each individual request may be presented with the suggestion that it is not significant enough to necessitate a change to the overall project schedule. The advice to the project leader is to learn to say no when it is appropriate to do so. When the change request seems justifiable, it is important to record it and to get agreement from the customer on the extra time required for implementing the change, no matter how small. It is also important to instruct the team not to accept change requests informally from the customer, because this can often happen unless the teams are specifically reminded not to allow it. All change requests, no matter how small, should be brought to the attention of the project manager and system architect.

9.3.6 Programmability

It is usually worth considering making some internal ASIC configuration settings programmable. This is most useful when it is difficult or almost impossible to calculate the optimal settings in advance. For example, in a datacomms chip using shared packet memory, it is usually necessary to buffer up a minimum amount of data before starting the transmission of a packet. This allows for some variable latency in accessing the memory for the next batch of data, due to the memory being in use by some other process at the instant that access is requested. The number of bytes that are built up in advance to allow for this access latency is commonly referred to as a *watermark level*. Excessive values for the watermark result in excessive latencies between initiating a packet transfer and the transfer actually starting on the transmission medium, whereas too low a level can result in data underruns. If there are many competing modules using the same shared memory for a variety of different reasons, it can be very difficult to predict the correct value in advance. By making the watermark programmable, it is possible to vary the value in simulations or prototype testing. This programmability or flexibility reduces the risk of having to redesign at a later stage.

Another typical situation where some programmability reduces risks is in the case of priority arbiters for access to common resources. The most common example of this, again, is a memory access arbiter, which may have to arbitrate between multiple different requesting sources where each source requires access for a variable number of cycles, depending on the conditions that prevail in the system at that instant. If it is difficult to predict all possible combinations of load conditions in advance, the level of risk will be reduced by making the arbitration algorithm programmable via register-driven settings. Then two or three different algorithms can be tried out, and chances are that at least one of them will meet the requirements.

Adding a degree of programmability is also appropriate when the design is intended to comply with a standard that has not yet been finally agreed on or ratified. Rather than wait until all outstanding technical issues in the standard have been finalized, it may be possible to implement several possible options selectable under register control. Although this results in greater gate count and a requirement for additional simulation and testing, it may also result in getting a product to market significantly faster than would otherwise have been the case.

It is worth mentioning, however, that programmability does not come for free. It always results in a larger gate count, and it requires more extensive simulation and testing. Designers should exercise prudence in deciding when it is appropriate to make implementations programmable.

9.3.7 Regular Design Reviews

As the design progresses, one of the most effective forms of risk management is to hold regular design reviews. Complex modules and modules designed by less experienced engineers require closer scrutiny than do other modules. In the case of less experienced engineers, watch out for the tendency to opt for overelaborate solutions.

Reviews should be held at many stages of the design. It is clearly advantageous to spot major mistakes as early as possible. Design reviews take time and effort, and must be properly planned. However, when properly conducted, it usually proves to be time well invested, and it reduces the risk of failure or delays.

9.3.8 Early Trial Synthesis

Early synthesis at the individual module level is an essential risk-reduction technique. This gives an early measure of how realizable the design is. It involves only minor effort from the designers if standard scripts are used that are accurate enough for early trial synthesis that they do not require fine-tuning or customization. If an early trial synthesis of this type fails to meet the timing requirements by a significant margin, it probably means that the architecture of the module or subsystem in question needs to be redesigned. In the worst-case scenario, it may mean that the entire chip architecture requires significant redesign. In either case, it clearly reduces the risk to the entire project to identify this as early as possible. The trial synthesis also gives an early indication of gate count in the design. If this is well above initial estimates, it may also be necessary to redo some or all of the architecture work. Early trial synthesis is, therefore, strongly recommended as a risk-reduction measure.

9.3.9 Early Trial Layouts

Early trial layouts are also an effective method of reducing risks by identifying layout problems as early as possible. As soon as the first draft of the complete or almost complete integrated chip netlist is available, it should be sent immediately to the vendor for trial layout, irrespective of whether it is fully tested. If there are layout difficulties, the technology library may have to be changed to a faster technology type or the architecture may have to be changed. Again, regardless of the solution, it is better to know about this type of problem at the earliest opportunity because it reduces the risk of meeting the target end dates by avoiding wasting further time proceeding down a path that is not going to achieve the required results.

9.4 Reducing the Risk of Design Bugs

The most effective way of detecting bugs is to simulate and test the design thoroughly. There are many ways of doing this that are discussed briefly in the following subsections. In all cases, an essential prerequisite of an effective testing strategy is to have a comprehensive test list that has been adequately reviewed to ensure that there are no major omissions.

Traditionally, software-based simulators have been the main tool used in testing designs. However, this approach is beginning to struggle in million-gate plus designs. The simulations simply take too long. As a consequence, alternative prototyping-based test approaches such as emulation, FPGA modeling and early risk sign-offs are now becoming increasingly popular. A further benefit of these alternative approaches is that they inevitably allow much greater test coverage. It may be possible to run several thousand tests each day on an emulator or FPGA, compared with less than 100 on a software simulator. The test coverage achievable in a given timeframe is, therefore, higher. It is also easier to simulate random timing events and random simultaneous input conditions with real hardware. The types of error conditions that these random load conditions produce are notoriously difficult to detect and model in software-based simulations.

9.4.1 Simulation

The normal software-based simulators play an extremely important role in ASIC debugging. Simulators which run faster give the opportunity to do more testing (Chapter 14, "Design Tools," discusses different simulation tools, such as cycle-based simulators). However, the top-level simulation phase is constrained in duration because it is always on the critical path. Therefore, the testing strategy should be considered carefully. With the benefit of some creative brainstorming, it may be possible to design many of the tests such that they provide maximum coverage in a minimum amount of run time. Testing the core functions of the chip should be given higher priority than testing optional features. If there is not enough time to do all the testing, at least the core functions will have been tested thoroughly.

Depending on the software architecture, it may be possible to use low-level Verilog or VHDL driver routines in ASIC simulations that model or are exact equivalents of the low-level software ASIC driver routines that form part of the final bundled product software. In fact, if the driver routines are written in C using simple constructs, it is possible to translate these automatically to VHDL or Verilog. By using the same or equivalent low-level driver routines in the ASIC simulations, this effectively debugs the software and, thus, reduces risks in the software development cycle.

In addition to playing a useful role in verifying the architecture at an early stage in the project, high-level modeling is also a useful technique to complement detailed, low-level simulations and ensure that the final netlist performs the correct function with the required performance. High-level modeling can be useful for creating reference input and output data to provide independent checking of the final netlist.

9.4.2 Emulation

Emulation is a method of modeling the design in hardware. It is usually based on FPGA-type technologies. Emulators can run VHDL or Verilog code multiple orders of magnitude faster than standard software-based simulators. For some designs, this is an excellent approach to identifying bugs in the design. Internal chip nodes can be made available for analysis, if identified in advance. However, emulation is not without design effort. The code must be compiled and fitted in the emulator, and a test printed circuit board (PCB) is often needed. Structures such as large memories can be difficult to emulate, and, therefore, these are typically put onto the PCB. Alternatively, a smaller memory is often sufficient to prove the functionality of the device. Unfortunately, emulation tools are not cheap. If this approach is used, the design route should be modified to make as much use of the tool as possible. Limitations on clock rate can make interfacing to other systems difficult. This can be addressed by sourcing or designing specialized test units capable of running at lower test speeds.

9.4.3 FPGAs

The principle of FPGA prototyping is similar to that of specialized ASIC emulation boxes. FPGAs are increasing in size and complexity and can be used to test a chip or modules within a chip very quickly. They can often run at full clock speeds, although not necessarily over the full operating environment range. However, provided that they can run at room temperature with a stable voltage supply, they can be used for functional verification. FPGAs provide a number of features, such as internal memories, and some even include embedded processors. The big advantage of FPGAs is that bugs can be identified, fixed and retested with very fast turnaround times. This helps eliminate bugs in the design quickly, with a consequent reduction in the risk to timescales. With larger designs, it is sometimes difficult to map the code into the FPGA while meeting the speed requirements. If the design is too large to fit in a single FPGA, it can be partitioned across more than one FPGA. EDA tool vendors now provide tools that try to optimize the partitioning automatically to achieve the best timing results.

FPGAs can be used in the initial production runs, provided that they meet the necessary timing requirements over the full environmental operating ranges. This can further reduce the product development risk because FPGA lead times are typically less than ASIC lead times since they can be ordered well in advance of design completion.

A special test PCB must be designed for the FPGA. The test board should be designed so that all FPGA I/O pins are easily accessible for oscilloscope or logic analyzer monitoring. Large FPGAs can cost several thousand dollars each, when ordered in small quantities. However, they are still significantly cheaper than custom emulation boxes, which can cost up to several million dollars.

9.4.4 Fast-Turnaround ASICs

Another effective approach to identifying bugs prior to tape-out is to use a fast turnaround ASIC. These are easy to produce and can provide turnaround times from second sign-off to silicon of the order of 1–5 days. There are many specialist fast-turnaround ASIC vendors; additionally, some of the mainstream ASIC vendors also offer a fast-turnaround service.

Fast-turnaround ASICs often lag mainstream technology, so sometimes the maximum required operating speed is not achievable or size limitations come into effect. For example, the amount of available internal RAM is typically smaller. However, this approach is good for identifying functional errors in the netlist. For identification of functional errors, the chip does not need to work over the full range of voltage and temperature. If the design is not pushing the limits of technology, the chip may well run at the environmental limits, and the entire design may fit in the chip. If this is the case, the project risk can be reduced by using the fast turnaround ASICs for the early production batches.

Additional resources are required to generate a netlist for the fast turnaround ASIC. At a minimum, the technology library will be different so that, even if there are no special cells, the design must be resynthesized or remapped. If any specific cells from the technology library are instantiated, rather than inferred in the code and a resynthesis, rather than a remapping, approach is being used, equivalent replacement cells must be found in the fast-turnaround library and the code changed to instantiate these instead. This is likely to be the case for any memory cell instances. The vendor can advise on the best approach to take for generating the new netlist.

The other important issue to discuss with the vendor is the sign-off criteria. The fast turnaround sign-off will require test vectors, although it is normal to settle for something less than the stringent 95% plus test coverage requirements associated with the final production ASICs. A smaller number of test vectors will require less effort to generate, but will add some risk to the chip (very basic vectors can be generated—with the assumption that most failures will be identified by simple vectors, such as IDDQ tests). Also, as the customer, you can waive the need to test the samples at the environmental extremes. Whatever the approach, it should be discussed, documented and signed by both parties well in advance of generating the netlist. Ensure that the agreement has been reviewed by a member of the technical team and not just a member of the purchasing department.

The fast-turnaround ASIC should have a request for quotation document (RFQ) that leads to a contract. Often, the vendor will allow a number of free respins of the chip. If so, ensure that the details are included in the contract and that there are no ambiguities. If the plan is to use the fast-turnaround ASIC for early production, this should be discussed with the vendor because the consequent increase in required volumes should result in better price deals. The delivery times of the production volumes should be discussed, because they may not be particularly fast. The ability to provide the required volumes for production quantities should also be assessed.

9.4.5 Early Sign-Off

Time to market is normally one of the key aspects of the development process. To reduce the risk of delays to the product launch date, the ASIC can be signed off early. Typically, top-level simulations take many weeks to complete. The gate-level simulations can be delayed when the VHDL/Verilog code requires many synthesis runs or changes to the code before the netlist is capable of running at the required speed at full environmental conditions. This adds risk to the final netlist sign-off date. An early sign-off of the netlist will reduce the risks of delays to the product launch date. The concept is similar to the emulation concept. Far more tests can be carried out in a much shorter time on real hardware. Additionally, the early sign-off will allow the software engineers access to the silicon earlier than would otherwise be the case, and PCB testing, including EMC and temperature testing, can be done with the early signed-off silicon.

The early sign-off should be based on a netlist that has completed some top-level RTL simulations and some gate-level simulations. There must be a reasonable degree of confidence that all the major functions are working to some extent. The project plan should schedule the most important simulations first. The early signed-off silicon may well contain minor bugs because not all the simulations will have been completed before tape-out. However, provided that the main functions are working, a significant amount of advance verification can be done. To allow for minor bugs with the early sign-off, the plan must allow for a subsequent respin of the ASIC, should it prove necessary. Therefore, extra cost and effort is incurred by doing an early sign-off. The ASIC respin should be done only when the early spin silicon has been tested.

9.5 Risks in Meeting ASIC Vendor Criteria

The ASIC vendor specifies a number of requirements for the ASIC and the test vectors in relation to sign-off of the netlist. If these requirements are not met, the vendor typically will not accept the netlist. Alternatively, in such a case, the vendor may accept the netlist only if the customer signs an agreement accepting responsibility for any manufacturing problems that arise as a result of the vendor's initial requirements not being met. Neither of these situations is desirable, and, therefore, as the project progresses, steps should be taken to avoid the risk of failing to meet the vendor's requirements when it comes to sign-off.

9.5.1 Power Consumption Issues

The power consumption of the ASIC should be calculated initially with presynthesis gate count estimates, then subsequently with the more accurate postsynthesis gate counts as these become available. Meeting a specific ASIC power consumption target can be an important deliverable for the project. The ASIC vendor will define maximum die temperatures for different packages. Power dissipation, therefore, has a direct impact on package type and, consequently, on ASIC cost. As synthesis progresses and the gate count varies (typically increasing), the power consumption of the chip should be recalculated. The temperature of the die has a significant impact on determining the performance of the chip. If the estimated ASIC power consumption exceeds target values, there are steps that can be taken to try to reduce it:

- Reduce the system clock speed. However, this is frequently not an option if external interface timings dictate the internal clock speeds or if the required performance or throughput simply cannot be achieved by using lower clock rates.
- Power down parts of the chip that do not come into play in the operating mode that is consuming too much power.
- Use gated clocks, which reduces the power consumed by flip-flops that are not being updated every system clock cycle.
- Consider using an alternative lower-power ASIC technology, based on lower operating voltages or library cells optimized for lower power consumption.
- Ultimately, the architecture may need to be changed to achieve the required power reduction.

Because some of these steps involve significant changes, it is important to review the power consumption on a regular basis to prevent wasting time continuing down a design path that is ultimately not going to achieve the required power goals.

If achieving a required reduction in power consumption to meet the constraints of working with a particular package type is not practical, it may be possible to change the package type. The package type contributes significantly to the temperature of the die (see below).

9.5.2 Package/Pin-Out

The package type is important in determining the cost of the ASIC and the ease of manufacturing in the final system. As pin count increases, the price typically rises, and the type of package changes from PLCC to PGA or BGA. The different types of packages have significantly different thermal resistances and maximum allowable die temperature. These thermal resistances affect the temperature of the die, and that, in turn, affects the maximum speed of the ASIC.

The ASIC vendor will specify a range of die sizes that can fit into the package types. It is important to ensure that the estimated size of the ASIC fits into the desired package with a comfortable margin. Often, the final size of the ASIC is bigger than expected. If the package has to change, this can be a major issue for the PCB design and can also have a significant impact on unit ASIC cost.

The ASIC vendor will want the pin-out defined as early as possible. An initial pin-out should be done early in the project to determine the required number of functional and power pins. The ASIC vendor will specify the maximum number of output pins that can be grouped between power pins. This design constraint is often referred to as the maximum number of simultaneously switching outputs (SSOs). The drive strength of the output buffers will significantly influence the number of outputs that can be grouped.

9.6 Summary

This chapter has identified some of the common risks in ASIC development and has suggested steps or approaches for minimizing these risks. The reader is further referred to Chapter 3, "A Quality Design Approach," and Chapter 4, "Tips and Guidelines," for complementary additional material. Probably the most important concept in reducing risk throughout all phases of the project is to hold regular design reviews. Make use of any available expertise, borrowing experts from other projects at appropriate times if they are not available for the full duration of the project. Making the correct high-level trade-offs (functionality versus cost versus performance versus timescales) during the initial project definition stage has a crucial impact on the subsequent levels of risk associated with the project. Consult widely and decide carefully during this initial stage. Finally, it is worth reiterating the importance of quality. Although quality design cannot guarantee a risk-free path, it can at least play a major part in significantly reducing the possible risks.

## CHAPTER 10

Dealing with the ASIC Vendor

10.1 Introduction

The ASIC vendor can be a real ally during the project. To achieve this, it is important to develop a good relationship and maintain good communications with the ASIC vendor throughout the project. Understanding the vendor's technology, process and tools will help in the smooth running of the project. If your company's business volume is sufficiently large, becoming a key customer of an ASIC vendor has numerous advantages: lower-cost ASICs, lower non recurring engineering (NRE) charges, earlier access to the latest technology, faster response time when dealing with problems and, most importantly, fast turnaround times for the silicon. Ideally, your company should develop relationships with a number of ASIC vendors. This reduces the dependence on one critical partner and allows the project manager the option to choose from a selection of ASIC vendors without increasing project risk greatly. Different vendors have different strengths and weaknesses in their technology offerings, and developing relationships with a number of vendors should allow a choice of technical solutions. Having good contacts with a number of vendors will also yield competitive pricing quotes for new ASICs.

10.2 Using the Vendor's Expertise

It is essential for a good project manager to keep abreast of ASIC technology developments. Failing to do so will result in less-than-optimal project decision making and, in extreme cases, the potential for complete project failure. ASIC vendors are, not surprisingly, a good source of information on ASIC trends. One way of keeping up to date with the latest developments is to organize technical update presentations at intervals of every 6 months or so from a number of key vendors. This assists the design team in determining what is feasible in the latest technologies and helps the project manager to concentrate on a small number of relevant vendors when a new ASIC development is being considered. The topics covered should be tailored to the particular type of ASICs the company develops but will typically include some of the following list:

- Latest-technology geometries and vendor's experience with these
- Maximum gate counts
- Availability of IP cores
- Availability of special cells (for example, analog front ends or phase locked loops)
- Memory types, densities and characteristics
- Maximum operating frequencies
- ASIC turnaround times
- Packaging options
- Methodology, design flow and tools
- Vendor services such as synthesis, floor planning and test vector generation

During the initial concept phase of the project, the ASIC vendor's input can be useful when analyzing design and architecture trade-offs, such as internal versus external memories, synchronous versus asynchronous memories, SRAM versus DRAM, maximum operating frequencies versus power limits, use of external off-the-shelf parts versus implementing these functions internally, etc. Understanding the ASIC technology, packaging and costs is necessary when defining the chip architecture.

10.3 Vendor Selection

Choosing the right ASIC vendor is a very important project decision. Ideally, the ASIC vendor will have been used during previous projects. Regardless of whether the vendor has been used before, it is necessary to evaluate its ability to deliver the best solution for the specific project in question. A formal request for quotation (RFQ) should be written that addresses all the technical aspects of the ASIC. The RFQ should also define commercial and production issues, and, therefore, it should contain input from a number of departments. The primary objective of the RFQ is to identify the ASIC vendor that can deliver the best solution for the proposed ASIC.

The RFQ should be sent to a number of vendors, and they should respond in writing, backed up by a presentation. The presentation gives the opportunity to meet the vendor face to face and to clarify any open issues. The RFQ and RFQ response, specify the commitments and deliverables of both parties, along with the development and production costs of the ASIC. The development charge (NRE) covers costs such as customer training and support, layout, chip mask making and initial samples.

The RFQ also defines the timescales for the vendor to complete specific tasks (for example, layout). Competing vendors tend to be more aggressive with such commitments while trying to close down a deal than they will be with any subsequent commitments that are not covered specifically during this period. It is, therefore, in the customer's interest to tie down as many specific issues as possible during the RFQ phase.

10.3.1 RFQ Details

The RFQ should cover the following topics:

10.3.1.1 Tool and Design Flow

The RFQ should contain a list of the tools that will be used during the project. The vendor's response should indicate whether the tools are compatible with its ASIC tool flow. The tools considered should include simulation tools, synthesis tools, test-insertion tools and test vector generation tools. The design flow should define how and in what format the data is transferred. For example, an FTP link can be useful for transferring design data, particularly where large databases are involved, because many email systems will break up large attachments into multiple smaller ones.

The vendor should also state which tools it is mandatory to run on the design in order to pass design rule checking. This is important because some ASIC vendors may specify tools with which the company has limited or no usage experience. This can add risk around sign-off time because the vendor's engineering team can identify problems that the project team cannot verify because it either does not have access to the tools or is not sufficiently experienced in using them. Additionally, for the same reasons, the design team may not be able to verify that changes it makes to the netlist will actually resolve the problems highlighted by the tool. A method of reducing this risk is for the project team to have access to and experience of using all the tools required well before ASIC sign-off. The ASIC vendor will also have experts in the tools who can assist the project team in learning to run the tools.

10.3.1.2 Technical Support

The strength and expertise of the local support engineers should be assessed. It is important that the local support engineers have experience with the design tools and have a strong relationship with the tool designers and tool experts at their development centers. The support engineers will be crucial during the critical sign-off stage. A good support team will reduce risks and keep the project on track. A weak support team may result in schedule slippage. The RFQ should ask for information on the support engineers for the project. What is their experience in general terms and with the current sign-off tools? It is also useful to understand what percentage of their time will be spent on the proposed project during various phases, but particularly around sign-off time.

10.3.1.3 Chip Definition

The RFQ should contain a top-level description of the proposed ASIC that includes the following items:

- Gate count: An estimate of the size of the ASIC in terms of number of gates. It should also explicitly state requirements for any compiled cells, such as memory cells, multipliers, etc. For memory cells, the specific configurations (e.g., 1 Kbyte x 8 bits), numbers of ports, operating speeds and other characteristics (e.g., synchronous versus asynchronous) should be specified.
- Target technology and operating voltage: This is the silicon technology (e.g., 0.25 micron 3.3 V technology).
- Package: The number of I/O pins and the type of package. This has a major impact on the cost of the device.
- Power estimation: This will be an approximate figure at the start of the project.
- Core cells or IP cores: Some vendors offer predesigned modules often referred to as *core cells* or *IP cores*, that can be used in the ASIC to reduce timescales and risks. These include microprocessors, DSP cores, analog functions such as PLLs, special I/O cells and many more. It is important that these be defined in the RFQ because some of these modules may not be available in all technologies. Using vendor core cells will often impact the cost. The cost can be a one-time payment or can be a royalty payment per device. If the payment is a one-time cost, the RFQ should state whether the payment covers minor changes to the core required for future upgrades to the device.

10.3.1.4 Configuration and Size Options

Ideally, the RFQ will contain a number of possible chip configurations. It is useful to request quotes for the ASIC with a number of different gate-count estimates.

Typically, the final ASIC will be larger than the original estimate. It is, therefore, useful to determine the production costs for a number of different gate-count estimates at the start of the project, when the vendors are competing against each other. Because larger dies may not fit into the desired package, it is useful to find out how close a given estimate is to a maximum gate-count for a given die. This information is important when assessing the risks associated with the project.

10.3.1.5 Chip Quantities

The production quantities will have a major impact on the price of the device. The production lead times and ramp-up times should also be specified in the RFQ.

10.3.1.6 Engineering Samples

The RFQ should also include engineering sample quantity requirements. These are normally expensive and quantities are limited. The initial devices are often manufactured with different equipment from those that are used in production. These lines of equipment have limited capacity and are expensive. The NRE usually covers the cost of the first five devices. Typically, there is a variety of types of engineering samples, each of which are available in different quantities and have different delivery times and costs.

10.3.1.7 ASIC Vendor Tasks

The RFQ should also specify the ASIC vendor's input throughout the project. This includes participation in regular meetings, provision of training and various other technical services.

- Progress meetings: The project leader should hold regular meetings with the ASIC vendor. The RFQ should state the frequency of these meetings and who is expected to attend. Initially, these meetings probably need to take place every 2 weeks, but should take place more frequently toward the end of the project. The meetings should identify and track the various issues that arise during the course of the project.
- Design and synthesis: The vendor's layout and application engineers need to meet the design team early on to discuss the design and synthesis approach. The design of the clock tree should be discussed. This should be done at the start of the design stage, when the ASIC architecture is defined but detailed design has not started in earnest. Once the synthesis approach is agreed on, a standard synthesis script can be made available early in the design phase of the project. For deep submicron designs, the synthesis and layout tasks must be more closely linked than was the case with previous-generation designs, and this area, therefore, requires particular attention.

- Training on the ASIC vendor's design flow and tools: The number of engineers requiring training should be stated. This training should be done during the design phase.
- Provision of a sample packages: If the ASIC uses a package that has not been used before in the company's manufacturing process, sample packages are useful to test the prototype production facilities. This reduces the risk of damaging scarce initial ASIC samples. The sample packages often have test silicon that allows solder joints to be tested.
- Trial layouts: It is useful to carry out a number of trial layouts prior to the final netlist. The early layouts give a good indication of how feasible the design is in layout terms. If a design proves difficult to lay out, it is clearly valuable to know this as soon as possible, so that various alternative approaches can be considered.

10.3.1.8 Miscellaneous RFQ information

Any other miscellaneous information relevant to the working relationship between the development team and the vendor should be included in the RFQ and RFQ response. For example, one important issue that arises with certain vendors is the shutdown of production facilities during holiday periods. If this coincides with the wrong phase of the development cycle, it can add cause significant delays. Additionally, if any of the support services have abnormal shutdown times or are not available during the normal working hours of the development team, due to international time zones, it is important to be aware of this.

10.3.2 Vendor Comparisons

The project leader should organize ASIC vendor presentations from each of the vendors under consideration to go through their RFQ responses. Relevant representatives of development, manufacturing and purchasing departments should be invited. It may also be appropriate to invite members of the senior management team if the project represents a major company investment or if there are long-term strategic vendor alliances under consideration. These presentations will allow any open issues to be answered. It is important during these meetings to assess whether a good relationship between the project team and the ASIC vendor can be developed and maintained. After the vendor presentations have taken place, further meetings can be held to compare the different ASIC vendors' quotations and to decide which vendor to choose.

In making the final vendor selection, a number of criteria should be considered. The cost will have a major impact on the decision. Cost includes production and NRE costs. However, cost is not the only consideration. Time to market is often critical, in which case, turnaround times for the first ASIC and any potential respins should be analyzed. A further important factor is the risk associated with opting for any particular ASIC vendor. How good is their technology, knowledge of tools, support experience, etc.? What about production issues: fab availability, production volume, delivery timescales, ability to ramp-up or ramp-down production volumes? Finally, strategic issues should be included in the decision. In some cases, the natural first choice may be excluded because close ties with another vendor might yield some longer-term advantage.

Prior to the meeting to select the vendor, it is useful for the project manager to talk informally to each of the participants to determine what they consider the major issues to be. These prior discussions will increase the chances of arriving at a decision quickly. Otherwise, given the number of different parties with an interest in this decision and the range of parameters involved, the selection process may drag on unnecessarily. This, in turn, may hold up decisions on solution architectures and, consequently, delay the start of the main development phase.

10.4 ASIC Vendor Services

ASIC vendors typically offer a range of services other than the principal service of laying out and manufacturing the chips. These vary from vendor to vendor but may include design, synthesis, test and floor-planning services. These services may be of interest because they can reduce risks when the project team does not have sufficient experience of particular tools, technology or functions. They can also bring in timescales if there is a resource shortage in the company team on the project in question.

However, using the vendor's services is not risk free. It is important to identify how much experience the vendor has in providing a given service before deciding to hand over any tasks. Ask the vendor how many times the service has been used by customers before and, ideally, ask it to identify other noncompetitor companies who would be willing to talk to the project team about their experiences of using these services.

A common service that ASIC vendors offer is test insertion and/or test vector generation. This task is heavily CPU-intensive and often requires more computing resources than are readily available in-house at the disposal of the design team. For this reason, it is often attractive to assign this task to the vendor. Typically, however, this service is available only for designs using a full-scan approach. Test insertion can affect the worst-case timing paths, and resolving this issue is often on the critical path of the project plan. The ASIC vendor should guarantee early test insertion to allow adequate time for timing verification and to provide a fault coverage figure. The availability of resources for final test insertion and test vector generation should be guaranteed, allowing for some deviation from target final netlist dates. The percentage fault coverage target should be agreed on early because this can affect the resource requirements of the project.

Another common vendor service is the provision of IP blocks or core cells, such as microprocessors, DSP cores or special I/O cells. The charge for providing these IP blocks should be clearly stated for both the current ASIC under development and for any future updates to the chip. Find out whether the IP block has been used successfully before. The vendor should also specify how to simulate and validate IP blocks after the silicon is available. An appropriate documentation package must also be provided, if necessary. For microprocessors and DSP functions, development tools such as compilers and emulators are important. Any such tools need to be available and stable, and the cost of using them agreed on. There also needs to be an approved mechanism for fixing or working around tool bugs in a timely manner.

10.5 Effect of the Vendor on Timescales

The project manager has direct control of the engineers who are working on the project. As the project progresses, in-house tasks and resources can be changed to address unexpected difficult problems or slips to the plan. However, the ASIC vendor also has tasks that are always on the critical path. It can be more of a challenge to influence these. The basic set of vendor tasks comprise layout, manufacturing of samples and manufacturing of production chips. The following section describes some of the ASIC vendor tasks and how these tasks interact with the design team's tasks. It also suggests methods of working with the vendor to improve timescales.

10.5.1 Layout

This is the process of mapping the gate-level netlist into a physical description of the layers that form the silicon chip. This task typically will be done at a central site, rather than the local support office. As such, the project leader will have less visibility of the work load of the layout group. Ideally, a senior layout engineer will be assigned to the project, with his or her time allocated to the project for both trial and real layouts.

The time taken to lay out the chip is difficult to estimate accurately. In the early project plan, it is advisable to be conservative. After processing the delivered netlist and carrying out trial placements and layouts, the ASIC vendor will provide a capacitance or wire-load file that can be converted into a standard delay format (SDF) file. The capacitance information should be fed back into the synthesis tool to resolve overloaded nets. Each ASIC cell is designed to drive a maximum capacitance, which includes the capacitance of the net connecting the output to the inputs of subsequent cells. Before layout, the synthesis tool uses an estimate of the capacitance of the net, based on an average figure. After layout, the net may turn out to be longer than the estimate, and this can increase the capacitance so that it exceeds the drive capability of the cell or increases the signal delay so that it breaks a timing requirement. Most synthesis tools can use the postlayout capacitance information to perform an incremental compile to the netlist to resolve any problems with overloaded nets. If the prelayout load estimates were sufficiently realistic, this should result in only minor changes to the netlist. A list of changes can then be given to the ASIC vendor, who will update the layout to resolve the overloaded nets. Each vendor has its own requirements when dealing with changes to the layout. Some require a list of the changes, whereas others need a new copy of the netlist.

Following layout, the design team should use the SDF information to carry out static timing analysis or to run a set of system simulations at minimum and maximum timings to verify that no timing paths are violating. Any problems identified here usually result in a resynthesis using the new SDF information to drive the synthesis and possibly also a new set of tighter constraints. If the violations cannot be resolved by resynthesis alone, some design changes may be required. Any redesign, of course, requires at least a partial resynthesis. In turn, any resynthesis will require some degree of layout change. This is an iterative process where layout results are fed back to the design team, which, in turn, uses these results to provide an improved netlist, if necessary. The iterations stop when all timing issues have been resolved. This process is often referred to as *achieving timing closure*.

Managing timescales during the iterative layout process is difficult but there are ways of attempting to control this phase. On the design team side, the use of reasonable timing margins in synthesis to allow for unfavorable additional layout timing delays will help to close the iterative loop. However, this needs to be approached cautiously because overcompensating in this way results in larger gate counts as the design is synthesized in a flatter structure with fewer layers of combinational logic. Furthermore, the design simply may not be feasible in a given technology if it is overconstrained. Additionally, the design team can improve the chances of success at layout by breaking up anticipated long timing paths with the liberal use of pipelining registers on critical paths and by registering all module outputs.

On the vendor side, there are also numerous steps that the project leader can take. First, if trial layouts have been negotiated and agreed on, these will give an early warning of potential layout problems and the need to redesign certain modules. They also ensure that the vendor's team is familiar with the tool flow and specific architectural features of the ASIC that will influence layout results, so that when the real layout commences, there is no getting-up-to-speed time required on these aspects. Second, if turnaround time for layouts has been discussed at the RFQ stage and commitments made, the vendor is obliged to try to adhere to these. Third, if the issue of the vendor providing sufficiently experienced layout resource has been agreed on up front, including allowing for the possibility of some delays by the design team in meeting the original target dates for the layout phase, it is difficult for the vendor to back down on such agreement.

10.5.2 Provision of Engineering Samples

Provision of engineering samples is another of the vendor's tasks. The best approach to improving timescales here is to negotiate for the best times possible during the RFQ phase, when the vendor is still trying to win the business. It is important also that samples of any previously unused packages be made available in advance for trial assembly purposes. A sufficient number of initial samples should be negotiated to allow for damage to samples in assembly of the initial prototypes. Finally, because the chips are often manufactured in a different part of the world than the location of the design team, the vendor and the project leader should ensure fast and efficient transit of the engineering samples. This includes checking that all customs paperwork is on order on both sides. It is frustrating for a design team that has gone the "extra mile" in signing off on a chip to hear that the samples have been sitting in a customs office or delivery company's warehouse for 5 days because the transit logistics were not properly organized.

10.5.3 Production Chips

The ASIC vendor should also make wafers without the metal layers. The die should include spare gates for small fixes. The extra wafers can be used to reduce turnaround times in case of quick fixes to the chip. They can also be used to have faster ramp-up times when the chip has been verified.

10.5.4 Liaison with the Vendor During the Project

Liaison with the vendor during the project typically takes place on a number of levels. First, on a formal level, there are regularly scheduled meetings between the project leader and the vendor's team. The project leader may choose to involve one or two of the design team's senior technical people in these meetings also. A recommended frequency for these meetings is once every 2 weeks during the early design stages, increasing to once weekly as sign-off approaches. However, the frequency will be related to the size and complexity of the ASIC, and it is up to the project leader and vendor to agree on a sensible meeting schedule. The following list suggests a set of topics that should be discussed and tracked at the formal meetings:

- Key milestones
- Deliverables
- Definition of compiled cells and IP cores (e.g., RAMS/multipliers/microprocessor modules, etc.)
- Trial netlists
- Pin-out
- Current netlist status
- Identification of critical signals
- First sign-off
- Sign-off documentation
- Parametric test vectors
- Functional test vectors
- Netlist change requests
- Second sign-off
- Manufacturing and delivery of samples
- Manufacturing of preproduction and production quantities
- Sign-off for production quantities
- Production quantities available
- Payments

Second, on a less formal level, there are ongoing technical information exchanges between the vendor's support engineers and the design team. These consist of impromptu phone calls and email exchanges and, if necessary, site visits also. It is advisable to channel design team inquiries at this level through one engineer on the design team. This prevents different designers from duplicating each other's mistakes and repeating the same questions to the vendor support team. The single point of contact can also track the queries and responses, and report on the effectiveness of these communications channels to the project leader for input to the formal meetings.

Finally, there are additional occasional contacts that take place throughout the project between the design team and the vendor's support team. Typically, this takes the form of on-site visits by the vendor's technical team to get specific tools up and running, to run process and tool training for members of the design team or to learn more specifics about the design.

10.6 Summary

This chapter has examined the relationship between the ASIC vendor and the project team. Starting with the advantages of developing relationships with more than one vendor (competitive quoting, reduced risk, wider range of technology solutions available), we then looked at the process of selecting the best vendor for the current project. The RFQ was described in detail. Remember that the vendors are competing against each other during this phase, and, consequently, this is the best time to tie down commitments from the vendor and to negotiate a favorable deal. Some vendor tasks and services were examined, and methods of improving timescales were suggested. The chapter finishes with a description of the communications channels between the vendor team and the design team during various stages of the project.

## CHAPTER 11

Motivation and People Management

11.1 Introduction

One of the prerequisites of running a successful project is creating and maintaining a successful team. It is important to remember that it is not only the project that needs to be managed—the team also requires management. If the team is not encouraged and led in a structured and considered way, the resulting team effort will fall short of its maximum potential. Running a successful project is not as simple as picking a group of intelligent people and handing them a project plan. They need to be capable of working effectively with others and communicating their ideas, they need to be capable of using the tools at their disposal to maximum effect, and they must be capable of sustaining a focused effort throughout the duration of the project. The team requires constant motivation, and individual team members require individual management approaches based on their skills, experience and personality traits.

There are numerous comprehensive texts available on people management and motivation. This chapter serves as an introduction to some of the common ideas presented in these texts. It also suggests methods and techniques for translating some of these ideas from abstract concepts into practical steps that are applicable to engineering development teams.

11.2 Managing Engineers with Different Experience Levels

An ASIC team will typically consist of a group of engineers with varying levels of experience, confidence and ability. The project manager should manage each individual in the most appropriate way to maximize the work produced by the team. It is useful to consider each individual engineer in terms of where he or she lie, in a matrix of developmental phases. The matrix in Figure 11-1 considers four distinct developmental phases in an engineer's development.

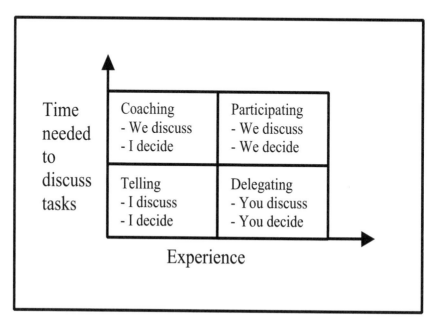

Figure 11-1   Developmental Phases

This figure serves as a simple but useful starting point in analyzing how to manage different members of the team. The project manager should understand the development phase of each member of the team and work toward moving him or her further along the development path. In each phase, the project manager interacts with the engineer in a different way. As engineers progress through the different phases, it is important that they understand that the expectations on them have changed and that there will be changes in the way their work is conducted. For example, in advancing through the various developmental phases, the amount of guidance and supervision from more senior team members and the project leader will be reduced.

Let's consider the matrix in more detail.

Initially, engineers have little experience and generally need to be told what to do by the project manager or system architect. This is the *TELLING* part of the matrix. This is the lowest developmental phase on the matrix, and once a new engineer has adapted to this basic operating mode, he or she should be moved quickly to the next phase for continued motivation.

As their experience increases, engineers should be encouraged to generate ideas for what should be done and how it should be done, instead of always relying on being told by someone else. The project manager or system architect will still take the actual decision on which idea to follow. This is the *COACHING* part of the matrix.

To improve motivation further, the engineer and project manager or system architect should jointly decide what is to be done. This significantly increases the level of responsibility placed on the engineer. This is the *PARTICIPATING* part of the matrix.

In the next phase of development, engineers who are sufficiently experienced and confident can be given a task with agreed-on timescales and then left to get on with the job and complete the task. This requires little management and some restraint on the part of the system architect or project manager. This is the *DELEGATING* part of the matrix. This is the most motivating phase for the engineers. It is important to realize, however, that leaving experienced team members to get on with the job themselves does not mean that their design work should not be reviewed and checked. Quite apart from the quality-checking aspect of design reviews, the reviews themselves can also act as a motivating factor. They are an opportunity for senior team members to showcase the quality of their work.

11.3 Maslow's Hierarchy of Needs

Figure 11-2 shows the hierarchy of needs as defined by Abraham Maslow.

In general, engineering team members will tend to be more concerned with the higher levels of the pyramid. Self-esteem and the respect of the other team members are important. The project manager should always show that he respects and values the ideas and views of the various members of the team.

11.3.1 Physiological Needs and Safety Needs

These are the most basic needs in the hierarchy. They concern providing a level of working environment, financial package and benefit package, such that employees and their families can live happily and securely. This is not really an engineering project issue. However, good project managers will try to ensure that the company properly rewards team members who perform well.

11.3.2 Social Contact

Social contact between team members and between engineers from different projects and departments acts as a positive motivating factor. Encouraging social com-

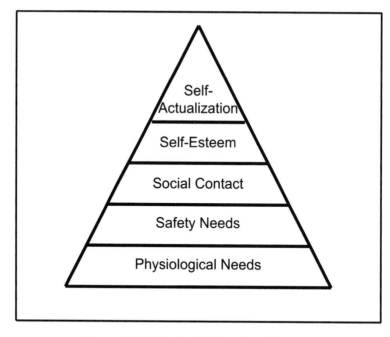

Figure 11-2   Maslow's Hierarchy of Needs

munication between engineers also helps working communication within the team. The project manager can encourage social contact by organizing lunchtime or evening events. These events could include sporting activities, visits to restaurants, social evenings and purpose-designed team-building events.

11.3.3 Self-Esteem

It is important that the engineers have confidence in their ability and that their contributions are recognized by the company and other team members. Respect from peers is very important and can be stimulated in many ways. A well-conducted design review for which the engineer is well prepared is a good opportunity to get visibility for their work. Other simple but effective steps in getting recognition for engineers are to encourage them to give technical presentations and to participate in company task forces and forums. For example, being asked to contribute to the definition of new or upgraded company procedures within a given field is recognition that the person so asked is regarded as highly competent within that field. The project manager can also guide engineers on how to perform during weekly meetings and brainstorming sessions to increase peer respect. The work that engineers are given is also an important factor in determining their self-esteem and motivation. Challenging work is one of the strongest motivating factors for engineers.

Engineers should have regular one-on-one meetings with the project manager in which they jointly discuss the project and the engineer's performance. These should be open discussions, and the project manager should endeavor to show that the company is committed to developing the engineer's career.

11.3.4 Self-Actualization

Self-actualization involves realizing one's potential for continued growth and individual development. Arguably, the individual must meet these needs entirely from within (hence the *self* in *self-actualization*), and, as such, there is little the project manager can do to assist. However, the project manager can at least attempt to ensure that a culture or environment is created wherein the individual will achieve self-actualization. Involving the senior team members in decision making and ensuring appropriate training for them will encourage the self-actualization process.

11.4 Getting to Know the Team

The team is a collection of individuals with potentially different views, interests, desires and approaches to work. Each member will have unique strengths, weaknesses and levels of experience. Spending time to get to know the individuals on the team will help the project manager to identify tasks that will be motivating for them.

The project manager relies on good communication with the team. Taking an occasional break from project talk to talk about some social topics often breaks down artificial barriers and provides a mechanism for better communication. Trying to appreciate the interests of the team members outside of the work environment shows that the project manager is interested in the person and not just the work.

It is also useful to have regular one-on-one meetings with the project team members. The meetings should be informal, with honest communication encouraged. The project manger can use the meeting to give feedback and to understand which tasks and roles would provide most motivation for each person. These meetings also provide a useful forum for the team members to give their views on the state of the project. They can identify any issues they have and highlight any interpersonal problems with other team members.

11.4.1 Different Personalities

Each team member has a different personality that affects the way he or she communicates. Let's consider the following two distinct personality types:

- an impulsive person who generally reacts instantly and wants things done *"Now!"*
- a reflective, analytical person who is more conservative and needs to think things through fully before responding to ideas or situations

It is interesting to consider how these two behave after having a good idea. The reactions are arguably exaggerated in the following text and assume extreme cases of each personality type, but this helps to clarify the example.

The impulsive person will immediately discuss the idea with others, solicit feedback and want others to accept the idea immediately and generate a plan. The ramifications and implications of the idea will not have been thought through before discussing it with others. The positive features of the impulsive type include the ability to motivate and generate enthusiasm in a team. They are very action-oriented and can encourage others to realize a plan. They will be happy to contribute to brainstorming sessions because they voice suggestions without the need for time to analyze the idea. However, in a crisis, the impulsive type will want action and typically ignore other people's views, suggestions or feelings. The resulting crisis action plan will not be fully analyzed, so there is a real danger that people will embark on tasks that will not solve the problem.

The reflective type will fully analyze the idea, decide whether the idea is appropriate and try to understand the implications. Only when they have had time to think through the ideas will they talk to others. The positive features of the reflective type are that they will stay calm during a crisis, they will be good at analyzing data, and they will carefully plan the solution to problems. During a crisis, their approach to solving the problem will generally be correct. However, they like to think through all the implications and understand all the data. Unfortunately, this takes time, and sometimes they will not make a decision when not all the data they would like to have is available.

It is important to understand how the different personality types will respond to other people approaching them with new ideas or questions. This affects how well the team members communicate with each other, which, in turn, can be an influencing factor in creating good team spirit.

11.4.2 Interacting with the Impulsive Type

The impulsive type will get caught up with a new idea and want to generate actions to make the idea work. They will not think through all the implications. The originator of an idea will be motivated by the enthusiasm of the impulsive type but may feel that they have not received any useful detailed analysis of the idea.

Consider the following two situations when the project manager and the impulsive type communicate: the first is when the manager approaches the impulsive type with a request, issue or idea, and the second is where the impulsive type approaches the project manager.

The first situation is relatively straightforward. The impulsive type will generally accept changes and new ideas easily. Care should be taken to ensure that they do not get carried away with the idea before proper analysis.

The manager should be more careful in the second situation, where the impulsive type initiates the exchange. They will want some agreement and will desire some actions to be generated. It is useful for the project manager to show enthusiasm about the idea and suggest some steps that will help to develop the idea further. Often, the new idea or issue will be uppermost in the priorities of the impulsive type. This can conflict with project deliverables. It is important to stress that project deliverables have the highest priority while not dismissing the issue or new idea. The project manager should explain the reasons for the priorities and persuade the impulsive type that the issue or new idea will be developed or dealt with in the future. A typical characteristic of the impulsive type is not to listen when under pressure, so the project manager must ensure that they do listen.

11.4.3 Interacting with the Reflective Type

For the reflective, analytical type, the potential problem situation is the reverse of that with the impulsive type. The reflective type will initially resist change and new ideas. It is important to prewarn them of changes or new ideas. They will immediately resist and often criticize the change or new idea. However, after time to reflect on the change or new idea, they can be fully supportive. They will also think through the implications and are, thus, a good source of feedback. The best approach to the situation is to introduce the issue or new idea and ask not for an immediate response but to take some time for analysis and agree to discuss it at a future date.

A typical characteristic of the reflective type is to delay decision making. The project manager should assist them with this process and, if necessary, take a level of responsibility for the decision.

11.5 Goal Setting

Goal setting is a classic way to motivate individuals. Ideally, the project plan should include some short-term, achievable yet challenging goals, thereby providing motivation early in the project. Often, complex projects will take many months of effort from the team. It is difficult to be fully motivated when working toward an end date that is many months in the future. By setting a number of short-term goals or milestones, the team will focus on short-term deadlines, rather than the final end date. When an individual's schedule slips by one day, he or she tends to judge the impact of the slip against the next deadline. If this deadline is months into the future or even the end of the project, the individual may become casual about slipping the day and also slipping subsequent days, convinced that there is lots of time to recover the slip. In contrast, a one-day slip will seem significant in the context of a short-term goal, and a greater effort will be made to pull back the day as soon as possible. It is important to remember that a project end date slips one day at a time. Another reason to focus the team on short-term goals is that the team members may not be on the critical path for the end date. If they are thinking only in terms of the end date, they will see that a slip in their deliverables does not affect the end date. However, in most projects, there are always unforeseen and, consequently, unplanned tasks that must be undertaken before a deliverable is complete. If those on the noncritical paths meet their timescales, they will have time to work on the initially unforeseen tasks.

Goal setting should adhere to the following rules. Goals should be:

- Defined

The goal must be clearly defined. The definition should include a detailed description of the goal, when the goal should be attained and what criteria will be used to define the completion of the goal.

- Achievable

The goal should be set such that it encourages hard work and is challenging, but the goal must be achievable. A missed goal will not motivate.

- Measurable

Goals should always include the measure that defines that the goal has been achieved.

11.6 Communicating Project Information

11.6.1 Project Meetings

Weekly project meetings should include a slot in which top-level project information is discussed. Individual designers often want this information so that they can see how their individual lower-level work fits into the overall development picture. The project manager should identify which top-level milestones the ASIC team is currently working toward. He or she should also identify the major achievements and milestones of the rest of the development team (for example, PCB design, manufacturing release, etc.). It is important to demonstrate that the ASIC development team is part of the wider product development team that is designing, manufacturing and marketing the entire product.

11.6.2 Marketing Information

Most team members are interested in and can be motivated by well-presented marketing information relevant to their project or line of work. Ideally, a marketing person who is capable of presenting the information in an exciting way should present this. As a guideline, a marketing information presentation should be updated every couple of months because markets and marketing expectations can change quite rapidly in technology sectors. Projections such as the expected run rate, anticipated market share, and expected profit are usually of interest to the development team and should be included in any such presentation.

11.6.3 Competitors' Products

The weekly meeting is a good opportunity occasionally to discuss competitors' products. This is particularly relevant when a product has just been released that is similar to the product being developed by the team. The morale of the team may be affected if members think they are working on products that could be beaten by competitors' products when released into the market. It is important to analyze the competitor's product prior to the meeting, paying special attention to features, cost and quality. It is useful to highlight areas in which the competitor's product is inferior to that being developed by the team. Often, the best analysis of a competitor's product is achieved by tapping the expertise of marketing, manufacturing and development in a combined analysis team. This is also a good opportunity to get members of each of these teams working together and can help break down interdepartmental barriers.

11.6.4 Highlighting the Importance of the Project to the Business

Periodically throughout the project, it is useful to request a member of the senior management team to attend part of the weekly meeting to express the importance of the project to the business. These senior management briefings, if properly delivered, will motivate the team. Everyone likes to think that his or her work is valued and important.

11.6.5 Show Enthusiasm and Have Fun

The project manager should try to encourage fun and enthusiasm in the project team. People tend to be more productive if there is room for a little fun and enjoyment. Otherwise, the project can become a burden. Team meetings can provide the opportunity for occasional lighthearted relief in the project. Care must be taken that the fun is not at the expense of individual team members. It is also useful if the project manager can convey enthusiasm for the project. This will stimulate enthusiasm in others, which, in turn, enhances motivation.

The ideal project manager will have a certain amount of personal charisma. This is needed to carry a large team through a difficult project. Project managers and team leaders should be selected not only on the basis of their technical ability but also on the basis of having the personality necessary to inspire a team.

11.7 Training

Provision of training not only provides a better-skilled project team but is usually also a motivating factor for recipients of the training. They feel valued because the company is investing money in them. Due to the pace of new technology developments, engineers, in particular, are keen to develop their skills constantly, and they place a high value on training. This is often quoted in job satisfaction surveys. Provision for training days should be included in the project plan because some will inevitably happen. The training can be split into three different areas: technical training, personal development training and product/design-specific training.

11.7.1 Technical Training

Technical training is normally highly valued by and motivating for engineering team members. This is particularly true when, for example, a new electronic design automation (EDA) tool is being introduced into the project. The confidence of the team members in achieving their goals and targets will be greatly increased with some specific training. However, it is important that the training be appropriate.

Training is often provided by external companies and arranged through the Human Resources Department. Courses should be evaluated by the participants to measure how appropriate and useful they prove to be. Irrelevant or poor-quality training can have a demotivating effect.

Courses on EDA tools can sometimes be designed and presented by the internal EDA Department. These can be excellent courses because they are focused specifically on the way in which tools are used within the company. They also provide a method of ensuring that company-specific configurations and usage styles for the tools in question are adopted by all engineers.

11.7.2 Personal Training

Personal training is very important for the long-term career progression of the team members. This type of training can be motivating and, in some cases, provides short-term benefits for the project. As with all types of training, the participants should evaluate the training course so that unsatisfactory courses are not used in the future. Team members are often working to tight timescales on the project. They will be demotivated if they take time out to attend a poor-quality course. In some cases, the project manager should allow team members to opt out of training courses if they are under serious timescale pressures because, in this case, even useful courses may demotivate them.

11.7.3 Product/Design-Specific Training

These are technical training courses that typically are created and run by internal company experts. These types of courses may cover background information on the product area (for example, explaining GSM phone systems) or the key features of the product being developed (the ASIC may be only one part in the product). They might explain design techniques that are used within the company (for example, low-power design techniques), or they might describe company standards, such as coding rules or company standard synthesis scripts.

These courses are usually well received and often satisfy designers' needs to have some information that forms a framework around their lower-level design. The courses can assist team members in delivering more suitable or better-tested design modules because they will understand the context in which their individual designs will be used in the greater system.

The courses also provide the secondary benefit of establishing the course presenter as an expert in the particular field. This can be a motivating factor for the presenter. It also serves to identify the presenter as a source of advice whom other team members can approach to discuss technical questions and concerns.

These courses and their preparation should be properly planned into the project activities because they may require significant preparation and participation time.

11.8 Summary

Projects are more likely to have a successful outcome if the project manager makes the effort to understand, motivate and develop each individual team member. There are some standard methods of getting to know the team members, which include identifying their personality traits and their experience levels. Understanding engineers and how they relate to others will allow the project manager to manage them in a way that will enhance their effectiveness and generate enthusiasm within the team. Motivating techniques such as goal setting, open team communications and the provision of effective training must all be used to maximum effect throughout the project to enhance the chance of success.

## CHAPTER 12

The Team

12.1 Introduction

A good team is central to the success of an ASIC project. High levels of individual and team performance are necessary to succeed in the current competitive global marketplace, where increasingly complex ASICs must be developed under ever more demanding time-to-market constraints. Although individual performance is very important, the team performance is vital to a successful project. The larger the team, the more important the team performance becomes.

If individual team members feel that the team as a whole is not working well, morale tends to decline. A consequent effect of this is that individual performance starts to decline. Ultimately, the project suffers, and, in the end, no one is happy.

In contrast, if the team members are working well together, motivation tends to be higher. Team members will be keen to support each other in their project activities and to work together to solve problems. This sort of cooperative spirit makes for a more enjoyable and productive working atmosphere, and boosts the prospects of running successful projects.

This chapter looks at some of the key roles in an ASIC project. Depending on the size of the ASIC, some of these may not be full-time activities for the duration of the project. However, they are all important roles, and it is important that the right people be chosen to fill them.

12.2 The Project Leader/Project Manager

The project leader or manager's role is clearly one of the most important roles in the project. It requires both technical and people management skills. These are discussed only briefly here in relatively broad terms because the detail is essentially contained in all the other chapters in the book.

12.2.1 Technical Skills

Regardless of how hands-off the management approach is, the ASIC project manager must have some level of knowledge of the technology involved. Ideally, he or she will have come from an ASIC development background. The level of knowledge required is that which is sufficient to enable the project manager to make a handful of key technical decisions throughout the project where options are available. For all of these options, he or she should consult the technical experts, but where there are differing views coming from the experts, the final decision is the responsibility of the project manager. Some of the key technical decision areas are:

- Which of several possible architectures should be selected?
- Is the design really feasible or are the technology risks too great?
- What simulation strategy should be adopted?
- What verification strategy should be adopted?

It is also worth bearing in mind that, as a general rule, the ASIC team members will have more confidence in and respect for a project manager who understands the technical issues fully than one who may well be a very good project leader but does not have a detailed understanding of the technology and is forced to rely all the time on advice from the team in making decisions.

Another good argument for having technically strong project leaders is that they can more accurately assess work loads and timescales. They can then more easily spot estimates from the team that are overly optimistic or overly pessimistic.

Project planning itself requires quite a degree of technical expertise. The project leader should be familiar with project planning techniques and tools, and capable of monitoring and tracking progress.

12.2.2 People- and Team-Management Skills

It is the role of the project manager to enable the team to perform to the best of its ability and to guide it through a quality approach in all aspects of the design. The project manager must gain the trust and respect of the team and, in return, must demonstrate his/her commitment to the team. The ideal project leader will have a degree of personal charisma because this is sometimes needed to inspire the team. A dull, lifeless technocrat is unlikely to inspire the team to great achievements!

It is important to convince the team that success will be reflected on each team member and not just the project leader or department manager. Each team member is an individual and needs to be managed in a unique way. It is useful to talk to each team member individually on a regular basis to identify any issues or concerns. Such issues may affect performance or add risk to the project.

The project manager should proactively promote team spirit. This is achieved when each team member considers himself or herself a valuable, respected member of the team and is confident that the other team members will help and encourage his or her work. A key component in generating team spirit is good, open communication. This can be achieved through regular, well-conducted team meetings and through occasional team social events. Possible social events range from hosting project team meals to organizing team sporting events or dedicated team-building events. Using a relatively small budget, compared with the cost of the entire project, these types of team-building occasions will improve team spirit enormously.

The project manager can also help team spirit by publicly encouraging and recognizing good team behavior. The regular team meetings present an opportunity to acknowledge and praise individuals who are acting positively in this regard. Inevitably, occasions arise when certain team members do or say things that adversely affect team spirit. If these situations persist, project managers should explain their concerns to the individuals in question and encourage them to modify their behavior. This is an important, if not pleasant, part of the role and requires delicate handling. Any such discussions should be held in private.

12.3 The Wider Team

ASIC engineers will form the core part of the team. However, several other engineering disciplines should have input to the design and specification of the chip. Software engineers, test engineers, manufacturing engineers and marketing engineers all have valuable expertise that can be tapped into. Some of this wider "peripheral" team should be encouraged to participate occasionally in team activities. This helps break down departmental barriers. It is especially important that the software engineers feel part of the team. As team members, they are more likely to support decisions, identify solutions to problems and be willing to volunteer assistance. (Quite often, a bug in an ASIC can be solved by a software workaround late in the development cycle.)

12.4 Key Roles Within the Team

There are many key activities in a project that merit special consideration and focused attention. It is useful to assign specific responsibility to some of the team members for these activities and, in certain cases, to create dedicated roles associated with them. The owners of these roles can then act as a point of contact and expertise for all team members in the relevant area. The project leader should create these key roles, explain them to the team and select suitable team members to fill the roles.

The following sections consider a number of such specific roles. These include the roles of system architect, tools expert, testbench engineer and team leader. The project plan should reflect the activities associated with these roles and allow an appropriate percentage of the selected team member's time to be allocated to the necessary activities.

The advantage of defining these roles and assigning team members to them is twofold: First, the individual engineers are more motivated with defined roles and challenging tasks, and second, problems in the role's areas of focus are likely to be identified earlier in the project.

Project leaders may well assign themselves some of the key roles in addition to the project management role itself. However, as team size grows, the project manager has less time to fill other roles. It is also important to understand that the project manager still has a responsibility to influence and steer particular tasks, even when another team member has been assigned to a key role.

12.4.1 System Architect

An experienced senior engineer should be appointed as the chief system architect, whose job it is to be responsible for the system definition and monitoring and management of system issues. This person works with the other members of the team to create an architecture that provides a good partition between hardware, ASIC and software functions. Ideally, the project manager should have considerable influence on the system architecture, because architectural decisions significantly affect the timescales and risks involved in the project. For example, the architecture could be based on reusing modules from existing chips. This would reduce risks and timescales but could affect the system architecture.

The system architect should be involved in all design reviews. The ability to conduct efficient reviews is, therefore, an important skill required by the system architect. The system architect should actively try to identify any conflicts and ambiguities in the specification and work to resolve them quickly. Although the role of system architect is not a full-time role throughout the entire duration of the project, there is an ongoing overhead associated with it, even after the initial intense burst of activity at the beginning of the project. The engineer filling this role should, therefore, be allotted an appropriate amount of time outside of his or her other design activities and responsibilities to fulfill this role properly. On small designs, the project leader may also act as the system architect.

12.4.2 Tools Expert

ASIC design using a hardware description language (HDL) is a complex task. Consequently, many of the tools used in ASIC design are complex and updated at frequent intervals. New technology or increases in gate counts often require a change of tools or a change in methodology. There are numerous significant processing tasks required at the ASIC top level that are best addressed under the guidance of a dedicated tools expert:

- Creation of the top-level netlist
- Test insertion and test vector generation
- Verification of top-level timing
- Running ASIC vendor tools on the netlist and test vectors

One or more engineers in the role of tools experts should be assigned these tasks. Time must be allotted well before the final netlist is due to check that the netlist and test vectors can be generated quickly and without errors. It is useful to test the logistics of the method of data exchange with the ASIC vendor. The normal method of transferring data for large netlists is via an FTP connection. The FTP link should be set up and tested well in advance of transferring the first real netlist.

Any tools issues should be highlighted as quickly as possible, because some tool bugs take a significant amount of time to resolve. The tools expert should actively track bugs in the tools and assess possible solutions. Where the tool provider cannot provide a solution, the tools expert must source alternative solutions.

This is a particularly important role. Toward the end of the project, the tools expert's tasks will be on the critical path. This role requires individuals who can work well under pressure and who can see tasks through to completion. Regular communications with the electronic design automation (EDA) tool and ASIC vendors is also part of the job description; therefore, good communication skills are also a requirement.

12.4.3 Testbench Engineers

System simulation is a key activity in the ASIC design flow. This is one of the stages in the project where it is easy to slip against the plan. System simulations typically take a significant time to run, and the testbench can be as complex as the ASIC. A good testbench will speed up simulation times, enable thorough testing of the ASIC and facilitate easy generation of functional test vectors. One or more engineers should be assigned the role of the testbench engineer. The role does not preclude its owners from also having design responsibility for a number of modules in the ASIC itself, but it is important not to underestimate the complexity of the testbench.

All or at least the majority of the ASIC engineers should be involved in the system simulations. The testbench engineer provides the framework for the system simulations, not the actual tests. They provide modules that generate input data (drivers) and modules that check output data (monitors). This topic is covered in greater detail in Chapter 5, "ASIC Simulation and Testbenches."

The role of testbench engineer requires good communication skills. Testbench engineers will have to explain the testbench framework to the other team members and provide some basic training and documentation on its usage. They also have to be able to respond promptly and positively to bugs found in the testbench. These tend to be discovered in intense bursts at the start of the system simulation phase, and, during this period, the testbench engineers may have to work under considerable pressure.

A testbench engineer must have or acquire good system knowledge. This provides an interesting challenge for prospective candidates for the role and is a good selling point for a job that attracts a certain amount of unjustified bad press. The role of testbench engineer is a high-profile role that will enhance its owner's visibility and, if carried out well, should also enhance their career significantly. This, too, is a good selling point for the role. It is particularly suitable for good, less experienced engineers, although it usually requires a certain amount of guidance from more senior team members.

12.4.4 Team Leaders

The terms *team leader*, *project leader* and *project manager* have been used in a fairly generic, broad sense in various sections of the book. In this section, a distinction is made between the roles of team leader and project leader or project manager. Perhaps it is best explained by using the term *subteam leader* because, in the context of this chapter, it refers to the leader of a team within the team.

When the ASIC team is large or the design is being worked on simultaneously in a number of different geographic sites, it is useful to have team (or subteam) leaders. In the split-site development example, the team leader is responsible for coordinating and managing the activities of the team on that site, under the guidance of the main ASIC team leader or project leader. In a single-site large team, the role of team leader might apply to a person charged with heading up a team of three or four engineers who develop a particular subsystem within the chip.

Good team leaders can be a real asset to the project leader. They will generate good ideas and provide feedback about project actions and suggestions before they are presented to the team. They can highlight problems with the project or morale of the team. They free up time for the main ASIC project manager because he/she no longer has to deal with each individual member of the entire team as frequently as is otherwise the case.

When people are assigned team leader roles, their opinions and views should be sought regularly. Major events and issues should not be presented to the team without discussion with the team leaders first. It is important that they contribute to the running of the project in a meaningful way; otherwise, the role is superfluous.

Team leaders should have good communication and management skills. It is very important that the project leader support the team leaders and encourage the team members to work cooperatively with them. However, the project leader should observe the relationships between the team members and the team leaders, and intervene occasionally, if necessary.

12.5 Summary

This chapter has emphasized the role of the team over the role of the individual in ASIC development projects. Highly talented individuals can make only limited contributions if they cannot work positively in a team context. For this reason, the project leader should nurture team spirit and encourage teamlike behavior from all the team members.

A number of key specialist roles in the ASIC development team have been described. The most important of these roles, aside from project leading itself, are ASIC architects, testbench engineers, tools experts and team leaders. The main tasks associated with the roles have been presented, and some of the skills and characteristics necessary to do these have been described.

## CHAPTER 13

Project Manager Skills

13.1 Introduction

The project manager can have a significant impact on the success of the project. To be truly successful, a project manager must combine technical awareness with strong interpersonal and administrative skills. Often, project managers are promoted from an ASIC design role, but the skills required for the positions are significantly different. The following subsections define three skills that are particularly important for the project manager role.

13.2 Running Meetings

The ability to run efficient meetings is a useful skill for project managers. Meetings are important for the project because they provide a forum for schedule tracking and team communication, and they are helpful for building team spirit and improving motivation. However, poorly run meetings are demotivating and counter-productive.

All meeting participants should have a clear understanding of the structure of the meetings. The most important meeting is the weekly progress meeting, and the format of this meeting should remain consistent from week to week. Many team members, if given the choice, would often prefer to stay at their desks doing work, rather than attend project meetings. There is a perception that meetings, or at least parts of meetings, are a waste of time. For this reason, the meetings should be kept short and to the point. Ideally, each meeting should have an agenda with an estimate of the time required for each item. The participants should be given adequate notice of meetings, and they should be provided with an agenda well in advance. Participants should be required to indicate their availability to attend in advance because some meetings will be pointless if key individuals cannot attend. It is better to reschedule a meeting than to find out after the meeting has started that some people cannot be present.

The chairperson of the meeting should do some preparation work. For each agenda item, any relevant information should be available prior to the meeting. If part of the agenda is to debate a particular issue, the chairperson should encourage the participants to think about the subject prior to the meeting. This will yield better results because the participants will have had time to form their ideas, present them in a coherent way and, therefore, obtain a consensus more quickly. Sometimes, it is useful to understand the views of the different participants prior to the meeting. This gives the opportunity to anticipate any contentious issues and therefore manage the meeting accordingly.

During the meeting, the chairperson should try to follow the agenda as planned to avoid excessive overruns. The chairperson should encourage decisions and ensure that they are not sidetracked. If a particular detailed issue is raised that involves only some of the participants, the discussion should be stopped and resumed at the end of the meeting, with only the relevant people present.

The project leader should follow each technical discussion, asking relevant questions but, wherever possible, allowing the team to reach a decision. Typically, discussions will result in an action or the agreement to adopt a particular approach. Sometimes, the team cannot reach a decision, and the project leader should give the final decision. However, where the issue has caused heated arguments, the decision may be delayed until after individual follow-up discussions.

Certain personalities can attempt to dominate meetings, whereas others do not actively participate. The chairperson should solicit views from all relevant parties. It is important that everyone feel that his or her views and ideas are worthy of consideration.

During the meeting, actions and decisions should be recorded. The actions should define the task, the person responsible and the date for task completion. It is useful to record the decisions and actions online if a portable PC is available because this saves translating notes at a later stage. At the end of the meeting, the actions and decisions should be summarized.

Following the meeting, minutes should be distributed as soon as possible.

13.3 Interviewing

A core role of the project manager is the formation of a team of engineers who can successfully deliver the project goals. The project manager is likely to be involved in the recruitment process, and the interview is the method used to select the correct people. Interviewing is a skill that must be mastered because inappropriate hiring decisions can turn out to be expensive mistakes.

Prior to the interview, the candidate's curriculum vitae should be studied and the interview plan formed. The plan should identify topics that can be discussed to determine the skills and knowledge of the interviewee. It should also indicate the time allotted to discuss each topic.

At the start of the meeting, the format of the interview should be explained to the candidate. It is a good idea to adopt a common format. For example:

- First, the interviewer gives a quick description of the type of work that is done in the department.
- Next, the candidate should be asked to explain his or her experience briefly. The idea is to relax the candidate and develop questions from the overview of experience. These questions should have been prepared in the interview planning stage.
- The questions should be phrased to encourage detailed responses rather than simple "yes/no" answers. For example, rather than asking "Do you use Synopsys Design Compiler for synthesis?" the question could be "What tools do you use in the design flow?"
- It is important to establish the candidate's depth of knowledge or experience. This is done by asking more detailed, probing questions. For example, "How long have you been using Design Compiler?" "What problems have you had with the tools?" "How did you solve your problems?" "Can you explain the content of the synthesis script?"
- Finally, the interviewer should describe the company to "sell" the qualities of the department and work. The interview is a two-way meeting. If the candidate is good, then he or she may get employment offers from a number of companies. It is important for the project manager to establish what the candidate's job expectations are. It is pointless to make an offer when the job opportunity does not meet the aspirations of the candidate because he or she may quickly become demotivated and leave, which is detrimental to the project and the project team.
- The interview should run to the allotted time. The project manager should ensure that all topics are covered, allowing the candidate the opportunity to ask questions.
- It is important that a number of people interview the candidate. The topics to be discussed by each interviewer should be agreed on beforehand, and a

review of the interview should be carried out as soon as possible after the event, while memories are still fresh.

13.4 Time Management

Time management is a skill that everyone should practice. Good time management is especially important for project managers because they may frequently be confronted with unpredictable issues that require immediate attention. In this role it is sometime difficult to allocate set times to complete specific work. There are some simple time management techniques that help to use time efficiently.

- The most important technique is to generate a list of tasks commonly referred to as a "to-do" list. This should be done at the start of the day and should include all tasks, whether short-term or long-term. Each task should be given an expected duration and a priority. A plan for the day should be formulated. It is also important to reserve times during the day to resolve issues with the team.

- As each task is completed, the task should be deleted from the list. At the end of the day, the list of completed tasks should be reviewed. It can be quite motivating to see that scheduled tasks have been completed.

- The task list should be reviewed every day or every few days, for long-term tasks. These tasks should be planned into the project manager's diary. If a task is not planned, it will not get completed on time. This approach ensures that deliverables are not produced in a panic.

- Project managers should identify that part of the day during which they tend to be most productive for concentrated work, such as analyzing plans and risks, generating documents, etc. This part of the day should be allocated for this type of concentrated work. If an issue arises during this allocated time, the project manager should try to delay instantly reacting to solve the problem and, instead, finish the detailed task at hand first. This is not always possible, but interruptions during the allocated time should be minimized. For example, the telephone could be set to voice mail during this time.

- Recording minutes directly onto a portable PC during meetings can increase the project manager's efficiency. It ensures that nothing is forgotten afterward, and saves the time that it would otherwise have taken to type it up at a later stage.

13.5 Summary

This chapter has introduced three essential project management skills: efficient running of meetings, interviewing skills, and time management. Although much of the material is pure common sense, it is nonetheless worthwhile referring to it on occasion, just to check that you are putting what you should know into practice.

## CHAPTER 14

Design Tools

14.1 Introduction

The quality of the design tools will have a major impact on the project timescales. Correct selection of tools can save huge amounts of effort and yield a high quality final product. Therefore, it is worthwhile to give considerable thought to the tools at the start of the project.

As the size and complexity of ASICs grow, new approaches have to be undertaken. All project leaders and senior engineers have a responsibility for current and future designs. The most successful project leader will understand the future direction of designs and tools, and should migrate the design flow to use new tools. Although the use of new tools is important and interesting, they pose risk to the design. Each new tool should be well tested before it becomes part of the critical path of a project. The project leader should be conservative and, unless a new tool will give a major advantage to the project, it should not be included as a "core" tool (core tools are those that must be used during the project). This means that timescales will be based on tools previously used in the company. However, in the interests of future designs, some new tools can be tested alongside existing tools. This provides potential benefits to current and future designs.

Clearly, when ASIC design is new to a company, existing tools are not available. In this case, it is best to limit the number of tools used and to use only tools that are well established in the market.

The following section describes the different tools that are available and can be used at the different stages of the project.

14.2 Hierarchy Tools

There are many tools that can be used to capture chip hierarchy diagrams. These tools facilitate a top-down approach to design. The output of some of these tools can be used to produce the structural code that connects lower-level modules together. They are used when defining the top-level architecture and when creating block diagrams for the submodules. Any selected tool should be easy to use and accessible to all team members. The entire team should be encouraged to generate hierarchy diagrams and to keep them up to date.

14.3 Input Tools

Input tools in this context are those tools that are used to generate source code. They fall into the category of low-level text editor tools and high-level automated tools.

14.3.1 Text Editors

There are many text editors available, and most tool vendors include links to particular editors. Many have extensions that recognize VHDL or Verilog syntax and can highlight identifiable language constructs. The more sophisticated editors will run commands (e.g., *compile* command) and automatically display the results in a split window.

14.3.2 High-Level Entry Tools

There is a growing number of tools that allow designers to enter their design graphically in the form of high-level flowcharts or state diagrams. These tools will automatically generate synthesizable source code. They are particularly good for certain applications, such as state-machine design, that are typically best described diagrammatically anyway. After a short period of usage, many designers prefer to use high-level entry tools over text entry. The general consensus is that timescales are not improved significantly but the quality of the final work is of a high standard that is easy to review and modify. The diagrams provide good automatic documentation, so the module is easier to reuse, and reuse timescales are reduced.

This type of tool helps less experienced engineers to generate synthesizable code. However, it does not do away with the need to understand the architectural implications of the schematic constructs chosen.

Some of the tools will also output C code, as well as the common Verilog or VHDL. This allows the chip code to be tested in the C environment with higher-level C application code.

Synthesizable code can also be automatically generated from state tables.

14.4 Code Analysis Tools

Like C Lint tools, Verilog and VHDL Lint tools can be used to analyze the quality of the source code. They are simple and quick to use and can identify problems prior to simulation and synthesis.

14.5 Revision Management Tools

Revision management tools or source code control tools are vital for controlling the VHDL and Verilog source code in ASIC designs. These tools store different versions of the code by identifying the differences in successive versions. This allows efficient storage of the different versions.

Such a tool can be used to produce a release of the code needed for an entire top-level simulation environment, which includes the ASIC code and the testbench code. Continued updates to the code can be done while allowing easy access to the version that was used to do the top-level testing of a previous design. This can be useful when debugging a problem with the silicon of previous designs.

Some of the revision management tools have complex features, allowing multiple versions of a module to be modified in parallel, then automatically merged back into a single piece of code. This type of feature should be used with care but it allows several engineers to update the module simultaneously. Some tools help with multisite projects, allowing automatic code updates between the sites.

14.6 Testbench and Validation Tools

The quality of the final silicon is highly dependent on the quality of the validation and simulation. If the validation approach is well thought out, the quality of the silicon will be dependent on the number of tests that can be gotten through by the sign-off date. Typically, projects operate under tight timescales, and, consequently, test time is limited. Good, fast tools will allow a sufficient number of tests to be executed within the limited validation time available. The following section highlights some of the tools that are available.

14.6.1 Testbench Tools

The testbench is sometimes one of the most complex and difficult modules written by the team. In recent years, a number of tools have become available that, for some designs, reduce the time required to create a test environment or create higher levels of confidence in the test coverage provided.

A number of vendors offers tools that create testbench models that range in size and complexity. At one level, they form a single component of a testbench to perform or check a specific function. A typical example of this is a bus model that can be programmed to check all types of bus accesses and to check automatically for correct operation. This can be especially useful because it provides independent checking of a part of the design that must be compliant to a standard specification. For nonstandard buses, models can be automatically generated from timing diagrams.

At the other end of the spectrum, there are tools available that create the entire testbench environment. These tools allow the project team to work at higher levels of abstraction, using specific testbench languages. Most of these tools will provide some fault coverage capabilities that analyze what percentage of the design has been tested. There are tools that check the tests against a specification and define how much of the specification has been tested. These can be particularly useful when multiple versions of the chip are planned, because the investment of effort to create the specification can be shared between multiple projects.

14.6.2 Code Coverage

Code coverage tools identify the percentage of the code that has been tested by a simulation run. These tools are quick and easy to run, and are useful to ensure that the design code has been fully exercised. They are particularly useful during module testing, where the size of the design is smaller and where 100% condition, branch and statement coverage is a realistic goal.

14.6.3 Standard Simulators

There are very many standard simulators available from a host of vendors. Independent reports are produced that give comparisons of speed of simulation and memory usage for the different tools. At the present time, it is recommended that all projects use a standard simulator, if only to do some small gate-level simulations. There are simulators for VHDL, Verilog or both VHDL and Verilog, and they often have the ability to include modules written in other languages, such as C.

There are three common types of simulators: interpreted, compiled and native compiled. An interpreted simulator reads each line of VHDL or Verilog and sequentially converts it while running the program. The compiled simulator converts the VHDL/Verilog to C before running the program using the compile and link command. This takes longer to prepare for the simulation, compared with the interpreted type, but will normally simulate faster. The native compiled simulator is the same as the compiled, but it converts directly into the host computer machine code, rather than C. These can be faster to compile, quicker to simulate and have lower memory requirements. Whatever simulator is chosen, it is important that it be suitable for register-transfer-level (RTL) code simulation and gate-level simulation when running with back-annotated delays that require significant amounts of host computer memory.

Most debugging of the design will be done with the standard simulator, so it is important that the user interface be extensive and easy to use. The simulator should have a waveform viewer and a source-level debugger. The project leader should ensure that each team member has received adequate training on how to use the simulation tools.

14.6.4 Cycle-Based Simulators

Cycle-based simulators will run many times faster than standard simulators because they simulate without high-precision timing information, updating signals only at cycle boundaries. Because the majority of the validation is functional testing, the high-precision timing simulations are not needed. Clearly, this type of simulator should not be used for gate-level simulations when the object is to identify timing problems. Timing problems can often be better identified with static-timing analysis tools.

Simulation speeds can be increased if the testbench is written specifically with cycle-based simulation in mind. If this requires a significant investment in man-effort, there is a risk that it may negate the benefits of the faster simulation. However, the investment will pay back immediately for large designs or in the longer term for future designs. Some analysis should be done before purchasing cycle-based simulation tools. In some cases, investing the money that would be spent on cycle-based simulators into faster host machines or larger memory can give similar increases in simulation performance.

14.6.5 Hardware Accelerators

Hardware accelerators are simulators with extra dedicated hardware that increases the speed of simulation by orders of magnitude. Accelerators are limited in the maximum size of designs that they can handle, so it is important that the size of the design matches the capacity of the accelerator. Accelerators are specialist pieces of hardware and can cost a significant amount of money. However, because they run at high speed, they enable very long simulation runs to be carried out that might otherwise not be feasible. This is useful for large, complex designs.

14.6.6 Emulation

With emulation, the design is synthesized and mapped into special hardware that is usually based on multiple FPGAs. External stimuli are applied to the hardware, and the design is run in real time or near real time. The external stimuli are applied via a printed circuit board (PCB), which must be designed and manufactured. This PCB can be a modified version of the product PCB, in which case, it has the extra benefit of partially proving the PCB design.

This approach is particularly good for validating designs that require very long simulation runs or stress testing with asynchronous events. The capacity of emulation tools is limited, and the cost of these tools is significant.

14.6.7 Cosimulation

Cosimulators are tools that allow the ASIC to be validated with significant amounts of embedded software with an accurate high-level model of the processor. This technique is very useful for finding problems that often arise between the hardware-software interface. Because the processor is cycle accurate, the simulation can prove that critical software timings, e.g., interrupt latency timings, are adequate for the design. Cosimulation allows software bugs to be identified and resolved before the silicon is manufactured.

14.6.8 Formal Verification

These tools convert gate-level netlists or RTL descriptions into representative mathematical models. Different netlists can be compared by comparing the mathematical models. The tool can be used in two ways: gate-level to gate-level comparisons, and RTL to gate-level comparisons.

The first, which is the most common, is to check that two gate-level netlists are functionally the same. This is useful when a module that has been proven is modified but the function remains the same (for instance, when changes are made to increase the speed of a module). Simulating such a module may take considerable time and may not identify actual differences between two gate-level netlists. Formal verification will compare the netlists quickly and thoroughly.

The second way to use formal verification is to compare the RTL description to the gate-level netlists. This is more difficult to achieve and takes more effort. However, this is an approach that can reduce risks and, when combined with static-timing analysis, eliminates the need for gate-level simulations. This would be particularly useful for reusable modules and will be an excellent approach in the future.

14.7 Synthesis Tools

The synthesis tool is one of the most important and one of the most complex tools used in the ASIC design flow. Its function is to convert the VHDL/Verilog code into a gate-level netlist. Scripts that define the required speed of the circuit, the size, target technology and design-specific constraints typically control the tool. These scripts are discussed in more detail in Chapter 6, "Synthesis."

Using the right synthesis tool from an established company is extremely important if ASIC design is new to the organization or the design is very complex. It is important to have good support from the tool vendor. Some synthesis vendors offer consultancy, turnkey design of a module or other additional services. Most synthesis tool vendors produce guidelines for coding techniques for synthesis and offer training on their tool. Some host Web sites that contain technical papers on synthesis issues and other useful information (for example, see the Synopsys Web site).

The synthesis tool converts the VHDL/Verilog code into gates of the target technology and produces reports of the speed, size, etc., of the design. Some tools produce information that can be used to guide the placement and routing layout tools. This usually yields a faster design in the final silicon because critical paths are placed first. The synthesis tool is also used to make modifications to the netlist, based on layout information "back-annotated" from the layout tools. This is typically used to fix overloaded nets and timing problems introduced by the layout. The latest style of synthesis tool takes account of layout information when optimizing the design. These can address the common problem of discrepancies between timings obtained prior to layout and those after layout. This issue will be increasingly important with the new deep submicron technologies, in which routing delays become dominant.

There are many types of synthesis tools. The most common type synthesizes RTL code. This is code that explicitly specifies all registers and the control signals that define the flow of data. Behavioral compilers synthesize code that has been written at a higher level of abstraction, so the control signals are implied. Behavioral compilers significantly increase the designer's productivity because the coding is done at a higher level. Some compilers can create optimized architectures for datapath-intensive modules, providing different implementations for adders and multipliers and allowing merging of arithmetic functions and automatic pipelining. Power compilers synthesize the netlist using techniques to reduce power consumption in the final silicon. This type of tool typically analyzes a simulation to determine the most frequently toggling register outputs. It then synthesizes the combinatorial logic in a way to reduce the number of nodes toggling and even creates gated clocks for registers (a clocked register consumes significant power, even if it the output does not toggle).

14.8 Static-Timing Analyzers

Static-timing analysis tools calculate the timing paths between nodes in the gate-level netlist. Based on technology library files and the gate-level netlist with capacitance and load information from the actual layout, the tool identifies worst-case timing paths and calculates hold violations. They will highlight all potential timing violations much faster than running gate-level simulations. Although these tools are very powerful, it is still recommended that at least a small number of gate-level simulations be run before sending the netlist for tape-out.

14.9 Summary

This chapter has provided a brief introduction to some of the more commonly used tool types in the ASIC design process. The reader is advised to scan regularly publications such as *EDN* magazine, which frequently run articles on many of these tool types and compare different vendors' offerings. Vendors' Web sites are also a useful source of information for those unfamiliar with electronic design automation (EDA) tools.

BIBLIOGRAPHY

Books

Reuse Methodology Manual (Second Edition), Michael Keating and Pierre Bricaud, Kluwer Academic Publishers

Application-Specific Integrated Circuits, Michael John Sebastian Smith, Addison Wesley

Deep-Submicron CMOS ICs, Harry Veendrick, Kluwer BedrijfsInformatie

Advanced ASIC Chip Synthesis, Himanshu Bhatnagar, Kluwer Academic Publishers

Magazine Articles and Supplements

"Handling Multiple Clock Domains in Scan Design." Samy Makar. *Integrated System Design*, November 1999. (www.isdmag.com)

"Ten Tips for Successful Scan Design." Ken Jaramillo and Subbu Meiyappan. *EDN Magazine*, February 17, 2000. (www.ednmag.com)

"IP/SOC/REUSE Special EDN Supplement." *EDN Magazine*, June 5, 2000 (www.ednmag.com)

ABOUT THE AUTHORS

NIGEL HORSPOOL has nearly 15 years experience as a designer and project manager on ASIC and other projects. He has worked extensively in image processing, networking, signal processing and computer graphics, and is now an independent consultant for the Philips International Technology Centre, Leuven, Belgium.

PETER GORMAN has spent over a decade working in FPGA and ASIC design, for companies such as 3Com and Philips—where he now serves as senior design consultant. He has held senior design engineering, project management and design team management roles in projects ranging from digital video and data networking to CD/DVD and USB.

INDEX

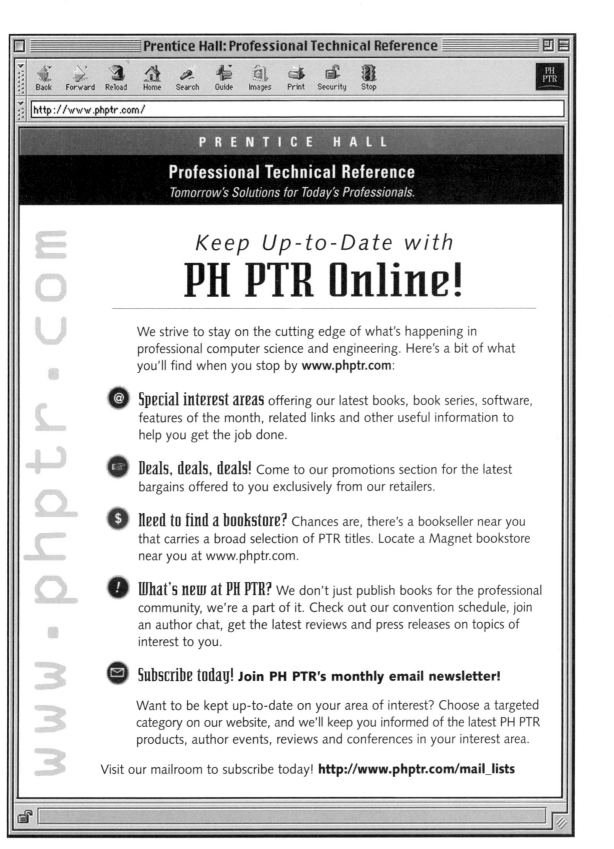